Quarterly Essay

CONTENTS

Quarterly Essay is published four times a year by Black Inc., an imprint of Schwartz Media Pty Ltd. Publisher: Morry Schwartz.

ISBN 978-1-86395-375-7 ISSN 1832-0953

Subscriptions – 1 year (4 issues): $49 within Australia incl. GST. Outside Australia $79.
2 years (8 issues): $95 within Australia incl. GST. Outside Australia $155.
Payment may be made by Mastercard, Visa or Bankcard, or by cheque made out to Schwartz Media. Payment includes postage and handling.

To subscribe, fill out and post the subscription card, or subscribe online at:

www.quarterlyessay.com

Correspondence and subscriptions should be addressed to the Editor at:

Black Inc. Level 5, 289 Flinders Lane
Melbourne VIC 3000 Australia
Phone: 61 3 9654 2000 / Fax: 61 3 9654 2290
Email:
quarterlyessay@blackincbooks.com (editorial)
subscribe@blackincbooks.com (subscriptions)

Editor: Chris Feik / Management: Sophy Williams
Publicity: Elisabeth Young / Design: Guy Mirabella
Production Co-ordinator: Caitlin Yates

PREFACE

As I began work on this essay, a recession in Australia seemed unthinkable and I expected that any suggestion of a near-term resources bust would be dismissed as implausible. In fact, the bust has come well in advance of publication, and recession now seems all but certain. As I finish writing, national economies are contracting the world over, stock markets are in freefall, the share value of Australian mining giants has fallen by two-thirds, most commodity prices have been cut in half, and the dollar has plummeted.

Yet, in a strange way, rather than challenging the Australian faith in digging and drilling, the bust seems to have fuelled it. The lucky feeling lingers on, except perhaps for those losing their jobs in quarries around the country. We remain confident that the "supercycle" will resume shortly and we credit our battered and bruised resources sector with shielding Australia from the full force of the economic blow suffered by other nations – at least for a time.

I am no opponent of the mining industry or the resources boom. Far from it. As I will discuss, mining has been as important to Australia's development as the sheep's back, and there is every reason to believe that it will remain so in future. My concern is with the "quarry vision" that has come to dominate our economic and political culture, in the form of vested interests and longstanding beliefs. This makes it very hard for the country to set a different course as circumstances demand.

In the era of global warming, such a change is urgently required. One of the core products of the Australian quarry is coal, and it turns out that coal is potentially lethal for the planet as we know it. In spite of the harm they cause, Australia has big plans to double its coal exports. We are already the world's leading coal trader, exporting 80 per cent of what we produce. This puts us up with Russia and the OPEC oil states as one of the world's leading carbon mules.

At the same time, Australia is bountiful in alternative forms of energy:

solar, wind, geothermal. But as it stands, our leaders are committed to expanding coal production, minimising greenhouse-emission cuts and shifting their cost elsewhere. In cosseting the coal industry, Australia is treating the poison in its economic well as if it were the jewel in the crown. Ultimately, this can only threaten our national interest and it must truly baffle outsiders.

How did the leaders of a modern, services-based economy come to decide that its future lay with a few carbon-laced industries? Why dig deeper just as the world decarbonises? Any explanation of why Australia has responded as it has, and why it is so difficult to imagine a different future, must begin, I believe, with the quarry.

Guy Pearse

QUARRY VISION

Coal, Climate Change and the End of the Resources Boom

Guy Pearse

It's our "natural competitive advantage endowed by providence," the "engine room of economic growth," "the backbone of our economy," or so we constantly hear from a political, business and media chorus. Debate rages about virtually everything else, but there's perfect harmony on the importance of the quarry. It's a given.

So it was nothing out of the ordinary when, in mid-2008, the federal Opposition leader of the day, Brendan Nelson, told a prime-time national television audience that it would be irresponsible to endanger our primary industries by introducing an emissions trading scheme as soon as 2010. After all, he argued, resources (mining, metals and energy) account for 37 per cent of Australia's economy. He had overstated the economic importance of these industries as a share of gross domestic product (GDP) threefold, but no one seemed to notice. Such is our collective quarry vision that no claim about the size or value of these industries much surprises us. From every direction, Australians are told that their current and future prosperity depends on what we dig, drill and smelt for the world. As a nation, we imagine the quarry's economic contribution is much as

Nelson described it. Some, no doubt, even imagine that coal alone is worth a third of GDP. We assume that many millions of our fellow Australians are employed by these industries, too. When it comes to business, the corporate diggers and drillers are our "world-class Olympic athletes." The rest of the economy is a mere sideshow; they are the main event.

That was the case during the resources boom and it's the case now, even as boom has turned to bust. The rapid industrialisation of China and India means that the supercycle must resume soon, we hope, and in that lies our salvation. We're assured that the developing world is hungry for Aussie coal, iron ore, liquefied natural gas (LNG) and other commodities; Australia is a mining colossus and "an energy superpower" that the world simply cannot do without. Furthermore, our exports are helping to clean up the planet: our LNG and uranium are replacing coal, and we're told that "Aussie know-how" means even our coal will soon be burnt cleanly.

Of course, while the boom lasted, some states were luckier than others in our "two-speed" economy. Parts of Western Australia and Queensland enjoyed bonanzas comparable to anything seen in the nineteenth century. The rest of us envied them, but only a little, for we all saw ourselves as winners. More Australians own shares directly and indirectly through superannuation funds than anywhere else in the world, and we imagine our stock market to be heavily "overweight" with mining and energy holdings. If our private stake wasn't enough, we also assume that government coffers across the country have been awash with inconceivably large tax and royalty cheques, and that without them everything from national parks management to the construction of new hospitals, schools and roads might grind to a halt.

The only limit on our good fortune, it seems, was our failure to anticipate quite *how* good it would be. Even more money would have trickled down if we'd just been a bit more prepared for the "skill shortages" and "infrastructure bottlenecks." The biggest problem was that we couldn't dig quickly enough. Had we known what was coming, we would have built more ports, railways and roads, and far more of us might have trained as geologists or engineers. Our dividends would have been even fatter, our retirements more luxurious, and even more of us would have been employed by resources companies. Of course, it was a shame you couldn't get a tradesman to come to your house while they were all off at the mines enjoying salaries beyond their wildest dreams, but we knew better than to get too greedy at such a generous buffet.

Now that the feast is finally over, we look forward confidently to the next one. Not for a moment do we foresee an inevitable collision between climate change and the quarry. Not for a moment do we pause and try to imagine a quarry without coal. The fact is, we are blinded by quarry vision.

THE RUSH THAT NEVER ENDED

To understand the sacred place of mining and related industries in Australia today, we need to think back. Mining is a large chunk of our history. Rushes for gold, silver, lead and tin opened much of the continent to European settlement and economic development in the late nineteenth and early twentieth centuries. More recently, coal, uranium, iron ore and alumina have dominated, along with the extraction of oil and gas. Though the price and relative importance of these commodities have varied dramatically over time, it has indeed been, as Geoffrey Blainey famously dubbed it, "the rush that never ended."

From the 1850s, countless inland communities were born out of mineral and hydrocarbon discoveries and the hysterical flood of money, blood, sweat and tears that followed. In some cases, as with Broken Hill, Kalgoorlie, Ballarat, Mount Isa and Charters Towers, today's visitor can readily imagine the heyday. Mining development shaped so much of Australia's built heritage and so many of our cherished landmarks. The banks and stock exchanges immediately conjure up the sense of confidence that prevailed. The density of pubs alone conveys the high stakes – the depth of the sorrows that needed drowning, the dizzy luck that needed toasting. The power of the rush is perhaps most evident in places such as Ravenswood, where scant but impressive remnants seem completely out of place in the ghostly and desolate surroundings. Ironically, the buildings still suggest permanence although the businesses named on the facades have long since vanished.

Whether the mining towns survived or were repossessed by the bush, they had an enduring legacy. For one thing, there is our transport infrastructure. As Blainey wrote, "the isolation of new mining fields carved lines of transport." Large mineral discoveries at Mount Isa and Kalgoorlie, for example, drove expensive new railways to some of the remotest parts of Australia and this drove extensive port infrastructure on the coast. "Gold finders spurred nearly every tropical port from Rockhampton to

Port Hedland, and south of the tropic every big port was enriched by the flow of metals." The extra investment, income and jobs gave governments a strong indirect interest. Mining licence revenue and royalties provided a more direct stake, as did government investment in mining and smelting companies. Without all that digging far, far away, cities around regional Australia would have evolved very differently indeed: imagine Townsville without silver, lead, zinc and copper from Queensland's north-west, Mackay without coal, Newcastle and Port Kembla without steel, Port Pirie without lead, Port Hedland without iron ore.

State capitals would have taken on a different complexion, too, especially Melbourne. In twenty years, the gold rushes led Victoria's population to soar from just under 12,000 people to more than half a million, most of whom eventually settled in Melbourne. Isolation gave the city an advantage. Lack of timely and reliable information made Australia's mines a risky proposition for foreign investors until the 1890s. Australian bankers filled the gap and for local lenders it became a core business. Melbourne became the pre-eminent economic centre in Australia, which it remained for many decades. For better or worse, the fortunes made from mining gave us the Melbourne Establishment.

At the other end of town, mining rushes drove large movements in labour, as people ventured to unthinkable locations chasing unimaginable wealth. The Wild West atmosphere of the Victorian gold rushes, soon replicated elsewhere, made mining towns a boot camp for industrial conflict. It gave rise to many of the Labor Party's most colourful leaders and characters. From a militant mining-union background William Spence rose to become "founding father" of the Australian Workers' Union and a long-serving federal Labor MP. Ted Theodore went from union organiser in Chillagoe to premier of Queensland and federal treasurer. Andrew Fisher rose from Scottish coal mines and Gympie gold mines to be a three-time Labor prime minister. The rapid influx of miners changed the political colour of electorates, giving the Labor Party a stake not merely in the interests of workers, but in the success of mining per se. As Chifley's decision to send the troops into the coal mines in the late 1940s amply demonstrates, mining was the scene of Labor's fiercest battles against communism. The passion with which mining interests are pursued today in Labor's parliamentary ranks, and the power that unions such as the AWU, CFMEU and AMWU wield, can be traced back to diggings around Australia.

At the other end of politics, the links are equally strong. Generations of mining-industry leaders had close ties to the world of conservative politics. Collins House, the "nerve centre of Australian mining," was instrumental in establishing Deakin's Liberal Party, financing the Nationalist Party and facilitating the rise of Sir Robert Menzies. The chairman of BHP, Harold Darling, wrote the Liberal Party's first industry policy, and Menzies' administration had a close relationship with the company. When war broke out, the managing director of BHP, Sir Essington Lewis, was effectively seconded to become director-general of munitions. He drove the development of the aircraft and automotive industries in Australia and was one of several larger-than-life mining bosses who saw themselves as patriotic visionaries and nation-builders. Many Australians shared their view that what was good for the company was good for the nation. The links between such men and non-Labor politics were strengthened through their involvement on the boards of think-tanks, particularly the Institute of Public Affairs, from the 1940s onwards. More recently, a succession of mining leaders have dominated cross-industry alliances, enabling them to become the voice of industry in Australia and to exert a disproportionate influence, especially on the Liberal Party.

From the beginning, these close relationships have often gone hand in hand with conflicts of interest, as public officials and politicians used their positions to profit from mining developments. The extra-legal activity ranged from police skimming mining-licence revenue right up to high political office-holders taking untoward advantage. Perhaps most famous was the "Mungana Affair," involving two Labor premiers of Queensland (William McCormack and Ted Theodore) who held interests in mining leases that were sold to government for considerable profit. The affair enmeshed the Scullin government in scandal and forced Theodore to stand down as federal treasurer. A generation later, in 1970, the other side of politics was left dishevelled by the "Comalco Affair," in which six Queensland Cabinet ministers received partly paid shares from which they stood to make an immediate paper profit as soon as the issued shares listed. Such situations have generally had little to do with ideology and more with mining companies hedging bets to ensure that business prospers no matter who is in government, and the less-than-proud tradition continues today. For every apparent scandal – from Warwick Parer's undeclared interests in the coal industry when he was the Howard government's resources and energy minister, to the big loans that Queensland Labor minister Gordon Nuttall accepted from coal magnate Ken Talbot – many more former ministers are quietly appointed to mining company boards, a parliamentary pension plan that guards against bad outcomes.

What with political scandal, larger-than-life characters, worker solidarity and the romance of striking it rich, mining has also contributed its share to Australian literature. Perhaps, as Russel Ward argues in *The Australian Legend*, this is because miners were better educated than the bush workers who preceded the rushes. Mining, and all its consequences, left its mark in bush ballads, and the *Bulletin* glorified the highs and lows – Lawson's "Eureka" comes to mind – so that mining joined agriculture as a fundamental aspect of the relationship that Australians have with the bush. In his *Golconda* trilogy of novels, Vance Palmer drew heavily on the rise of Mount Isa and on Ted Theodore for its central character.

Other items of Australia's cultural furniture – immigration, multiculturalism and xenophobia – were profoundly shaped by mining, too. The gold rushes of the mid-1850s peopled the country more dramatically than anything before or since, trebling Australia's population. Mining rushes sowed the seeds of multiculturalism, attracting large numbers of migrants to remote towns in Australia. The flood of Cornish copper miners into South Australia in the 1860s had an enduring effect on the make-up of communities such as Kadina and Burra. Elsewhere, Germans,

Scandinavians and Californians came in numbers. The influx of Chinese miners sparked xenophobia. The Chinese successfully picked over locations already mined and abandoned by whites, so were largely unaffected by the regulations effectively banning them from new finds for the first couple of years. White resentment led to rioting and violence – as at Lambing Flat – which went unpunished in spite of overwhelming evidence. Longstanding anti-Chinese sentiment on the mine-fields, pushed along by the economic downturn of the 1890s, provided the catalyst for the White Australia Policy, which had bipartisan support for most of the century that followed. Russel Ward even argued that racism may well be the *only* major addition to the Australian character for which we have the mining rushes to thank – most of the other traits were already well established by pastoral workers.

Mining exploration also brought Europeans into contact with Aboriginal people, with predictable results. In many cases, Aborigines defended their territory with vigour. Where Chinese miners were not banned from new digging, they generally knew better than to act as cultural icebreakers in an unfamiliar land. Many white diggers had no such qualms and, dreaming of a gilt future, found themselves speared. Go to Charters Towers Museum and peruse the selection of flint spearheads, mainly removed from unsuspecting white men, and you'll see what I mean. The retribution against Aboriginal attacks was often disproportionate and indiscriminate. Aboriginal Australians were summarily executed by rogue policemen whose recollections suggest the slayings meant little more to them than good exercise for a new rifle.

It's relatively easy to see the connection between mining and our physical and cultural surroundings, but our psychological attachment is more obscure. Faith in mineral and energy commodities is such an important part of the national psyche that TV newsreaders routinely assume the nation will not sleep without knowing the closing BHP and Rio Tinto share prices. Clinging to what we see as the safety of the coast, perhaps many of us are vicariously satisfied to know that someone, somewhere, has the upper hand and is making this vast, harsh, unfamiliar land pay. Perhaps mining feels more "real" than the intangible services most of us provide – just as the rush-era buildings seem more solid, authentic and dignified than the rushed buildings of today. Some may even like the idea that we're still "giving it" to the Chinese after all these years – making them pay anything but "mate's rates" for Aussie resources. There's certainly an element of patriotism in knowing that the brands and companies we link to national wellbeing are doing well: as the "Big Australian" goes – many think – so goes Australia. If the miners falter, surely other firms will get pulled down in their wake – Qantas, Holden, you name it. And perhaps, as with so much else, we like to hear we are world-beaters, "punching above our weight."

Or maybe the need to know all is well in the quarry stems from our punter mentality, which is directly linked to the mining heritage that so many Australian families have, whether they realise it or not. The vast majority of the half-million Australians who arrived in the years following the gold rushes of the 1850s headed straight for the mines. When almost all returned to live in the cities, they didn't stop trying their luck. Instead, people from all walks of life bet their financial future on hundreds of mining ventures. While gambling was frowned upon, even the church turned a blind eye to mining-share speculation. Mining didn't bring gambling to Australia, but it's become emblematic of this nation of punters. Many of us are still betting on shares like BHP and Rio Tinto, directly or through our superannuation. Even for those without a stake, the knowledge that all is well in the quarry helps us to feel that we still live in the Lucky Country.

And therein lies the problem. We misunderstand what Donald Horne meant: the intense irony with which he diagnosed our condition. Australia, wrote Horne, was a "lucky country run mainly by second-rate people" who failed to take account of their surroundings and ended up being caught out again and again. Writing of and in the Menzian "golden age," when the nation took the post-war boom as a given, Horne made various recommendations on how Australia should make its own luck. One was to become something more than a country relying on what we could dig up and ship out. Before his death in 2005, he bemoaned the "relentless propaganda about economic 'reform'" as a distraction from what really mattered. Our political leaders had removed tariff walls, floated the dollar and established an independent reserve bank – but they had not moved beyond the quarry. The changes, while necessary, were "concerned mainly with undoing some of the old controls and supports, not thinking about the conditions of producing something new." In colonial times, mining was one of our jobs as a colony, said Horne, but it was "beyond belief" that after all this reform there was still so much emphasis on "sending minerals to the 'tiger economies' so that they can develop more of the smart stuff." We were not yet the "clever country" we imagined.

As a nation, we ignore Horne's warning. We also fail to appreciate that if there is a common theme in the history of mining in Australia, it is that nothing lasts – bust follows boom, the rush always moves on to something new, somewhere new. It is not continuous, wrote George Seddon, "but more like herpes, breaking out in different ways in different places." No commodity or mine is indispensable – not even the biggest – and history's field is littered with those who pretended otherwise.

One hundred years ago, farming and mining accounted for more than 75 per cent of the Australian economy – today they account for around 11 per cent. Mining, metals and energy resources are not quite the economic security blanket they once were. As John Edwards, a former senior economic adviser to the Keating government and now the chief economist at HSBC Bank Australia, wrote in 2006:

> Mining is very valuable to Australia, but after decades of important discoveries in coal, iron ore, natural gas, copper, lead, zinc and uranium, the entire industry accounts for only 5 per cent of GDP. This is little different to its share of national output thirty years ago. It employs just 1 per cent of the workforce – half the share it employed twenty years ago. It is highly profitable, but it is mostly overseas owned, so most of the after-tax profit is sent offshore.

The recent commodities boom gave Australia its best terms of trade in more than thirty years, doubling mining's share of GDP in 2007–08 to around 8 per cent. Australia was awash with cash. Even so, at the height of the boom, as a share of our economy mining was less than half as important as car manufacturing was to the United States in its 1950s heyday. Factor in minerals processing, metals production, services to mining and the fossil-energy sector, and the Australian quarry all up accounts for less than 15 per cent of GDP. It's a very significant contribution, but much smaller than many Australians would imagine. By contrast, services such as retail and wholesale trade, finance and insurance, construction, health, education, business and personal services, and recreation generate more than 75 per cent of GDP. Rather than producing tangible goods, we are, as a sugar-cane farmer once sneered to me, "a nation of back-scratchers."

The place where mining and energy – especially coal – truly dominate is in the basket of goods that we export to the world. Coal, our largest single export, by itself accounts for around 10–20 per cent of our export income. As such, it is an important source of foreign currency, if a surprisingly small fraction of overall GDP: around 3–4 per cent.

What about quarry jobs? In recent years, thanks to high commodity prices, many people flocked to the mines, some induced by government subsidies worth tens of thousands of dollars to encourage workers into remote locations. However, thousands of these jobs have been cut since commodity prices fell, and there were never anything like as many as we imagine. Even at the height of the recent boom, mining accounted for only about 1.3 per cent of employment.

In Australia today, more people work in restaurants and cafes than in mines. Bunnings employs almost one-and-a-half times as many people as the entire aluminium industry. We might imagine that mining has a vast "knock-on" effect in creating jobs, yet even with services to mining, oil and gas production, mineral processing and metals production counted, the quarry and its closest dependants provide fewer than one job in twenty in Australia.

As the ships dispatch our mineral and energy treasures across the globe, we imagine *Australian* companies cashing in on the boom. Firms such as BHP Billiton and Rio Tinto own most of the iron ore, coal and other commodities – and we assume Australians own these firms. When we think of the power stations fired here with that coal, we also imagine the owner to be an Australian government or company. And we assume that most of the factories processing minerals or energy here are, in the main, locally owned. With each upward tick in commodity prices, we picture the "Big Australian" getting even bigger, and smaller local rivals steadily moving up towards the heavyweight division. We also imagine that the Aussie firms running our quarry are focused on Australia – that we are as important to them as we think they are to us. We are wrong on all counts.

Australia's mining companies and energy producers are majority foreign-owned, so most of their profits go offshore. BHP Billiton no longer publicly refers to itself as the "Big Australian" because it's no longer

Australian-owned. Though domiciled in Melbourne – largely as a lingering political inconvenience – it is foreign-owned and operated, as are Rio Tinto, Xstrata and Anglo Coal, the three other biggest players in Australia's coal industry. Foreign companies are even more dominant in LNG, one of Australia's fastest growing commodity exports.

The coal-fired electricity and gas produced for our domestic use are also much less Australian than we think. British and Chinese interests own most of the electricity produced with brown coal in the La Trobe Valley, and foreign firms are lining up to buy electricity assets from the NSW government. Downstream processing industries, such as aluminium, lead, zinc and steel, are largely foreign-owned. Australian-sounding businesses are anything but. For example, Cement Australia is owned by British, Swiss and Mexican interests, Queensland Gas is owned by British Gas, Queensland Alumina is owned by British and Russian companies, and Japan-Australia LNG Pty Ltd is a joint venture not between Japanese and Australian companies, but between two Japanese companies.

Part of quarry vision is a tendency to overstate our importance to the rest of the world. Commentators will often call us the largest "producer" of coal (false) rather than the largest "exporter" (true), glossing over the big difference between the two. Even with the boom-driven growth of late, China is expected to meet over 90 per cent of its own demand for coal for decades to come. So while our status as China's biggest coal supplier will become significant in future, as I'll discuss later, right now the claim is akin to highlighting Mauritius as Australia's biggest supplier of refined sugar. It merely distorts perceptions. Many Australians are led to imagine that the coal-fired power stations popping up like weeds across China are fuelled with Australian coal. Yet we provide less than 1 per cent of China's coal, and most of that goes to steel mills rather than power stations.

In a similar way, it was not long ago that John Howard declared Australia to be worthy of "energy superpower" status, a term enthusiastically embraced by the Rudd government. Yet when the phrase was coined,

Australia was making less money from our energy exports than we paid for energy imports – we were a net energy importer by value. Overall, we account for some 2.5 per cent of the world's primary energy supply and depend far more on foreign oil than the rest of the world does on our energy exports – hardly the definition of an energy superpower. Our resources are globally significant, but less so than we are encouraged to think.

Because we tend to accept uncritically that mining and energy are the engine room of Australian prosperity, we presume that these industries pay a large share of our taxes – especially the coal-producing firms. The recent resources boom roughly coincided with almost annual income-tax cuts, and, egged on by the media, we have intuitively linked the two. We assume that Queensland and New South Wales are riding a huge wave of coal royalty revenue. However, the state budget papers put things in context. In Queensland in a normal year, coal royalties account for about a billion dollars, or 3 per cent of revenue. Thanks mainly to a spike in coal prices that is now well and truly over, the Queensland government expects coal royalties to provide 9 per cent of operating revenue (some $3.2 billion) in 2008–09. But after this, such royalties will plummet. As for New South Wales, its coal royalties doubled this year to $1.5 billion, but that is still only 3.1 per cent of revenue. In Western Australia, mining royalties have also increased by about a third to make a much bigger contribution than in other states (some 17 per cent of total revenue). The big difference, however, is that the royalties are based on iron ore and LNG, not coal. The situation is similar at the federal level, where, between company tax and resource rents from oil and gas production, the resources sector contribution was much bigger during the boom but still a small proportion of the overall total. All up, the mining industry claims to pay $18 billion in state and federal taxes; it's impressive, perhaps three to four dollars in every hundred, but hardly the magic pudding we might expect.

And if we're to judge objectively what the quarry contributes, we really ought to subtract the various subsidies paid out by governments. Mining, metals and energy industries enjoy billions of dollars in tax breaks, fuel excise rebates, cheap electricity contracts, royalty holidays and infrastructure. The taxpayer provides some $9 billion annually in energy and transport subsidies: while much of this goes on fringe-benefits tax breaks, which encourage us to drive to work, plenty also helps to raise profits in the resources sector. On top of the tax breaks, government spends large one-off sums on infrastructure and project incentives. In Victoria, for instance, state and federal governments are jointly spending over $150 million to help build a new brown-coal-fired power station. In Queensland, the state government is subsidising "clean coal" and also spending $5.4 billion on a Coal Transport Infrastructure Investment Program. In New South Wales, half a billion dollars in federal money is being spent in the Hunter Valley alone to improve railway infrastructure to increase coal exports. The federal government is also spending $500 million on the National Low Emissions Coal Initiative. The resources sector is a very significant taxpayer, but one way or another, much of the money that the quarry industries contribute returns by a different route.

Beyond their role in public finances, we assume that various holes in the ground around Australia sustain our personal savings and underwrite our retirement. After all, the Australian Securities Exchange (ASX) is laden with mining shares – hundreds of mining companies have their own separate listing in the financial pages. Yet, important as the sector is, it's not what it used to be. In 1992, resources was the biggest sector in the ASX by market capitalisation – bigger than manufacturing, financial and other services; by 2006 it was the smallest, accounting for roughly 22 per cent of the total. While the value of resources stocks soared during the boom, it also fell fastest when the bust came. The key point is that our superannuation investments are less reliant on resources than many of us think, because they are deliberately diversified to minimise long-term risk. With Australian equities making up approximately a third of Australian super funds, roughly 90 per cent of most retirement income comes from something other than resources stocks.

Given all the benefits of the quarry, most of us are conditioned to view it as an unambiguously good thing: we thank the boom for supporting property prices and inflating the Aussie dollar to make overseas travel that little bit cheaper. But a wealth of resources can be a mixed economic blessing: mining ties up large quantities of capital that might generate more jobs in other sectors. Because the investments take decades to deliver their returns, the money is tied up for a long time. Rapid rises in commodity prices increase the value of what we dig and drill from the ground, but they have various side effects: they increase inflation, so interest rates go up to keep inflation in check; this attracts more investment in the Australian dollar, which rises and thereby reduces the competitiveness of everything from manufacturing to agriculture and tourism. This cycle digs us ever deeper into the quarry by compromising our capacity to pay our way in the world by other means.

As well as obscuring reality, quarry vision blinds us to history and its biggest lesson: nations squander the proceeds of booms with monotonous regularity. This has occurred so often that an abundance of resources has come to be seen by many economists as a guarantee of macro-economic volatility, even as a curse rather than a blessing – "Dutch Disease," some call it. As well as inflating currencies to the point that countries become uncompetitive, resource-rich nations tend to be excessively optimistic about the likely duration of boom conditions. This leads them to ignore or underestimate the impacts of inevitable busts. Even where it's clear that the boom commodity will run out, countries fail to invest in physical and financial infrastructure to help ride out the tough times that will surely come. From Saudi Arabia to Argentina to Nauru, we see the same trend. There are rare exceptions, such as Norway: since 1967, it has put oil revenue into a future fund now as big as the country's GDP to help sustain its social safety net once the wells run dry.

Australia has not learned that lesson. History told us that the billions of extra tax dollars generated by high commodity prices were temporary, yet almost all of the windfall was spent on new liabilities. The quarry did its bit, but our quarry vision let us down. As John Garnaut recently wrote: "the country voted for tax cuts, bought new cars, rebuilt homes and piled on more debts as if the China-driven resources boom would last forever." Our elected leaders did much the same. Where the tax proceeds of the boom were used to build infrastructure, it was often for the purpose of making it easier to dig up more commodities, as if the bust would never come. The consequences are now sinking in, as the global recession – particularly the downturn in China – blows a hole in federal tax revenue just as the government needs money to stimulate the economy and minimise deficits.

Ordinarily, we might get away with this. So what if mining is not the source of unending prosperity, so what if our stake is smaller than we think? Let the economists and technocrats argue about the shape of our economy and how best to use our resources in the national interest. She'll be right, mate.

Except that she won't – not when it comes to the largest product of our quarry.

We might not love a coal-burnt economy, but our inability to imagine something different is digging us deeper into danger as the immensity of the climate crisis hits home.

The health of the global climate system is inextricably linked with atmospheric concentrations of greenhouse gases. Of these, carbon dioxide is the most significant. CO_2 concentrations have ranged between 190 and 280 parts per million (ppm) for the last 650,000 years, but in the past 250 years they have risen sharply to more than 386ppm. The UN's Intergovernmental Panel on Climate Change, the largest global collaboration of scientists ever seen, is now more than 90 per cent confident that emissions caused by humans are predominantly responsible for the clear warming trend. And the overwhelming scientific consensus is that fossil-fuel burning, land clearing and agriculture are driving concentrations of greenhouse gases even higher; if nothing changes, they will ascend to well over 1000ppm this century. Climate scientists believe this would have catastrophic impacts, including an average 6 degrees Celsius increase in temperature. Even at current levels, we are seeing hitherto unimaginable developments, including the loss of most summer Arctic sea ice and the opening up of both the Northwest and Northeast Passages.

Returning CO_2 concentrations to the 300–350ppm level recommended by climate scientists requires cutting emissions beyond zero by mid-century. Industrial emissions will need to be all but eliminated, deforestation reversed and agricultural practices changed, such that the net impact of humans is ultimately to "draw down" carbon from the atmosphere rather than adding to it. Even then, 350ppm will take well over a century to achieve, and we will need to keep drawing down carbon to return the climate system to safe territory. No matter what we do now, some severe impacts are unavoidable: rising temperatures and sea levels will increase damage from cyclones, storm surges, bushfires, flood and drought. If we aim merely to stabilise greenhouse-gas concentrations at their current level, scientists agree that some catastrophic tipping points are a coin toss; if we stabilise emissions at a higher level, they are odds on.

These possibilities include the loss of the Greenland ice sheet or large sections of Antarctic land and sea ice, the release of methane trapped in permafrost and seabeds, and a cessation in the conveyor belt of ocean currents that redistribute heat between the poles and the equator. There is mounting evidence to suggest that these phenomena are either underway already or far too close for comfort. Some of the world's leading scientists believe we are already dangerously close to the circumstances at the end of the last interglacial period, when ice-sheet disintegration caused sea-level rises of four metres a century. The environmental, economic and social costs of adapting to the impacts that are already unavoidable are likely to be huge; and as atmospheric greenhouse-gas concentrations rise, these costs escalate. According to Sir Nicholas Stern's 2006 assessment for the UK government, the cost of adapting to climate change could be twenty times the cost of reducing emissions to avoid the worst impacts.

Australia is more exposed than most industrialised countries. Since 1950, temperatures have increased across the country, while rainfall has declined by around a third in the most populous regions. On current emission trajectories, we will see a rise in temperature of 5–6.3 degrees by the end of the century. Irrigated agriculture in the Murray-Darling Basin would be all but eliminated. Even if the world stabilises emissions at 550ppm, the frequency of extreme fire-weather is likely to at least double, four-fifths of Kakadu wetlands would be inundated by seawater, and the Great Barrier Reef as we know it would be gone. The severity of cyclonic winds would also increase, adding to the many hundreds of thousands of homes already threatened by sea-level rise, storm surges and other extreme weather events. So Australia has a big stake in the success of global efforts to combat climate change.

Simply to keep greenhouse-gas concentrations at their current level, global emissions will need to peak by around 2015 and fall by 50 per cent before 2050. If this happens, concentrations will temporarily "overshoot" before stabilising. The world is acting to achieve this, but all too slowly. Time and again, national leaders come together to say they will "seriously consider" and "work towards" a collective non-binding "aspirational target" of halving emissions by mid-century. Like a shoal of anchovies, dazzling but desperate for comfort in a group, they deliberately avoid talk of what the target implies for individual countries. Contrary to what is often said in the greenhouse debate to make advocacy simpler, no specific cut by any particular country is essential to the achievement of a global target. It can be achieved in an infinite variety of ways – the burden sharing is ultimately a political decision. That said, since emissions won't peak in most developing countries for decades, it's hard not to conclude that a 350ppm target can only be achieved if developed countries as a group make 80–90 per cent reductions by mid-century. The morals are as compelling as the maths. Developed countries are responsible for close to 75 per cent of the emissions caused by fossil-fuel burning since 1850 (more if we consider that Western consumption drives a large share of developing-country emissions), and per-capita emissions in China and India are minute by Western standards.

In theory, an effective global response to climate change is possible without Australia making deep cuts. In practice, the political consequences for Australia would be severe. We have the highest per-capita emissions in the developed world and we were one of a few countries granted an increased emissions target under the Kyoto Protocol. (And as the former environment minister Robert Hill noted, this is unlikely to happen again.) The world already cuts us plenty of slack. Whatever might be negotiated in Copenhagen in late 2009, a few things seem certain. All countries will face steadily more stringent constraints on industrial and agricultural emissions, and there will be a much greater focus on reducing deforestation. Developing countries will have to keep emissions well below the "business as usual" levels projected for 2020 and make absolute cuts soon thereafter. In developed countries, emissions will need to peak in the next five to eight years, with rapid absolute cuts well under way by

2020. Greenhouse pollution will carry a price that rises as constraints tighten and the cheapest emission cuts are made. Australia's heavy reliance on carbon-intensive coal and foreign oil leaves us exposed both to the onset of carbon pricing and to the reality that oil production is peaking.

While arguments rage over when the world will completely run out of cheap conventional oil, production has already peaked in more than fifty countries. It's generally agreed that "peak oil" globally will happen within five to ten years, if it has not already occurred. Some top-10 producers, such as Russia, have peaked recently and the ranks of oil exporters are thinning as countries keep more of their own supply. Meanwhile, the number of cars in the world is spiralling upward. A decade ago about one in 1000 people in China and India owned cars; today it's closer to one in fifty.

The global financial meltdown has suppressed oil prices temporarily, but in the longer term dwindling supply and mushrooming demand will ensure that prices resume their upward march. Australia is becoming more reliant on foreign oil just as supply becomes less reliable and more expensive. As recently as 1999, Australia was a net exporter of oil – today we are only 60 per cent self-sufficient. Quarry vision leads some to think liquefied coal is the answer to peak oil, a local path to energy security that can also be clean if carbon capture and storage (CCS) is used in production. In fact, were CCS to be used, the tailpipe and other emissions released would be equivalent to that of the fuels being replaced. Far from killing two birds with one stone, a home-grown problem replaces an imported one.

As oil and carbon prices rise, so too does the cost of producing very many goods and services. The quadrupling of oil prices seen in recent years has provided a sharp taste of what's to come. As oil became more expensive, biofuels became more profitable. More land and existing crops were diverted from food markets to biofuels, and reduced supply made food more expensive. In the court of public opinion, biofuel mandates were blamed, but the bigger factor – rising oil prices – was largely overlooked. So was the lesson: we need to find less oil-intensive ways to produce food *and* provide transport – not just to save money, but because that's where the new commercial opportunities lie. With the CSIRO warning us to prepare for petrol prices of up to $8 a litre within ten years, the lesson will become harder and more costly to miss sooner than many of us realise.

The most dramatic impacts will probably not be on the actual products and services we consume but on how they're provided. Most things are not inherently oil- or carbon- intensive – aluminium can be produced with renewable electricity, just as large cars can run on it. Until recently, however, business has been able to take it for granted that oil is cheap and plentiful, and greenhouse pollution is costless. Over the next decade, as that presumed competitive advantage evaporates, it will alter the relative competitiveness of different ways of producing and delivering goods and services.

The formidable challenges posed by priced carbon and peak oil are not beyond Australia. In spite of our dependence on foreign oil and coal-fired electricity, our overall economy is not especially energy-intensive. The vast majority of GDP and employment comes from industries that don't generate large quantities of greenhouse gas. Before the recent resources boom, ten industries responsible for 37 per cent of our emissions accounted for only 4 per cent of national production, 3 per cent of employment and 15 per cent of exports. Commodity-boom volatility aside, the most consistent growth in recent decades has been in sectors that are not especially vulnerable to either oil or carbon prices. Many industries even in the resources sector are far less emission-intensive than one might assume: LNG and iron ore, for example. Much of Australia's quarry remains extremely valuable in a carbon-constrained world. And since Australia has some of the world's best renewable-energy resources – the potential capacity for solar and geothermal electricity production in Australia is almost without parallel – we are relatively well placed to reduce our exposure to rising oil and carbon prices. Even with petrol priced at $8 per litre and carbon costing hundreds of dollars a tonne bearing down, we can take this in our stride if we act soon.

Not one credible piece of economic research suggests that making deep cuts in emissions by 2050 would cause even a temporary recession, let alone "crash" the economy, or "cut GDP," or send energy prices spiralling, or cause whole industries to shut down or flee our shores. Every serious study of the costs finds that deep cuts would *delay* the trebling of the economy and doubling of real wages by a few years at most later this century. The same analysis finds that acting sooner generates about a quarter of a million jobs more than would delaying, and many of the steps that reduce our exposure to carbon prices save rather than cost money.

However, you'd know none of this from the apocalyptic language that dominates the political debate. Rather than say that "halving Australia's emissions by 2050 would result in GDP growing 247 per cent rather than 281 per cent," as the government's own research bureau, ABARE, projected, John Howard used to say that it would "lead to a 10 per cent fall in GDP." The impression given is that the economy would shrink as compared with today. That's what happened during the Great Depression in Australia: real GDP shrank by 11 per cent between 1926–27 and 1930–31, and in 1932 unemployment hit 29 per cent. It took years for the country to recover and this is precisely the impression that so many dissemblers try to convey about emission cuts. The National Party's leader in the Senate, Barnaby Joyce, went so far as to claim that for Australia to more than halve its emissions by mid-century, either half the population would have to be exiled or we would have to accept impoverishment and living standards equivalent to those of Chad. "The ETS [emissions trading scheme] will have the impact of a heart attack," said his colleague Ron Boswell. Political insiders from both major parties have routinely warned against greenhouse policies that are "suicidal," that "strangle" or "wreck" the economy or "cut its throat." It's reminiscent of now laughable Howard-era claims that the Kyoto Protocol was the "most serious challenge to our sovereignty since the Japanese fleet entered the Coral Sea on 3 May 1942"; that "ratifying Kyoto would mean the closing down of all agriculture," "total power blackouts three days a week," and the banning of all passenger vehicles and light trucks.

The persistent notion that emission cuts will wreck the economy has provided Australia with the motive for delay, and good luck has provided the means. Under the Kyoto Protocol, the baseline year against which progress is measured is 1990. As fortune would have it, this coincided with the height of land clearing in Australia. Because forests absorb carbon dioxide, a reduced rate of land clearing means lower "net emissions" – 92 million tonnes less annually, according to the Department of Climate Change. This, along with another 21 million tonnes saved annually through expanded forest plantations since 1990, enables Australia to say it remains "on track to meet its Kyoto target," even though our actual greenhouse pollution – emissions excluding land-use change and forestry – has increased by more than a third in the intervening years. By relying almost totally on land-clearing cuts and forestry to achieve the target, Australia has avoided any serious action so far. In effect, the warnings that business and consumers should receive about the looming impact of rising oil and carbon prices are being intercepted and shredded by government to protect industries we wrongly believe are our economic backbone and future.

If we took these warnings seriously, we would shred something else: our decades-old coal-fired export strategy that to this day no government will admit is flawed.

A FOSSILISED GAME PLAN

The oil shocks of the 1970s accelerated international demand for coal. Soon afterwards, the Australian economy was turned outward: tariff protections were removed, the currency was floated, foreign investment restrictions were relaxed. Previously cosseted businesses did it tough, but trade liberalisation opened up new markets for our exporters and brightened prospects for the mining industry. Senior officials in the departments of trade and primary industries and energy persuaded the Hawke Labor government that coal was one of the keys to Australia's international competitiveness.

Other developed countries relied on protectionism to attract energy-intensive investments; we would use cheap coal. What Australia lost in manufacturing, it would gain in commodity exports and new investment in minerals processing, metal smelting and the like. This approach came to drive energy and trade policy. In the 1970s, the emissions intensity of Australia's energy supply was similar to the OECD average, but that would change as Australian reliance on coal-fired electricity intensified. All of this occurred just as global warming and peak oil were first emerging as serious political concerns internationally. Rather than revisit their coal-fired agenda, some Australian government officials buried the threats, even removing a draft chapter on climate change from a federal energy-policy blueprint. Recalling the late 1980s when this occurred, one long-time lobbyist told me that it happened because:

> senior public servants perceived it as their patriotic duty to prevent the coal industry from being undermined by an untoward focus on something that, in their thinking, was a load of cobblers.

Since that time, Australia has been determined not to let mounting facts interfere with a good story. In another interview, a senior energy policy adviser in the Keating government described the scene by the early 1990s:

> at the stage when we were talking about energy-market reform, breaking up the electricity market and reforming it, the view was we had to drive energy prices down and consumption up. [Q: Consumption up?!] Well, I mean so we would attract energy-intensive industries and therefore increase consumption. Yes, basically make Australia the homeland for footloose capital that required cheap energy – aluminium and so forth. And therefore we expected to see increased consumption of energy because that was our comparative advantage. When we went through the whole reform process, there was an attempt [by others] to get in there that there had to be a lot of fuel switching and greenhouse [considerations] and [that] prices should actually reflect carbon and all that sort of stuff. That was effectively removed by Keating.

Paul Keating recently confirmed his concern about protecting the coal-fired strategy, saying:

> I was around when Kyoto started. But I was always very cautious about the idea that Australian energy, which was part and parcel of our internationally traded products, coal, ingot aluminium, which is congealed electricity, that we were not putting cost burdens on them.

John Howard was equally set on this approach, both blocking Robert Hill's proposed greenhouse trigger for federal environmental approvals and overriding Cabinet to block emissions trading. Many of those who decided not to let environmental considerations get in the way bet that climate change and peak oil were ill-founded and overblown. "We were all sceptics," said the Keating policy adviser. Because of their wager, cheap coal came to be seen as Australia's "natural competitive advantage," dominant export and major selling point to multinational investors. We effectively chose to become the greenhouse ghetto of the developed world. Each advance in our coal-fired development has made emission cuts harder: a new aluminium smelter powered with coal, for example, is the emissions equivalent of adding a million cars to our roads; if we attract four new aluminium smelters, the greenhouse benefit we might gain from putting solar hot-water systems on every house in the country would be erased.

Instead we are stuck with one of the dirtiest electricity systems in the world at the worst possible time. With every upward tick in the carbon price, our cheap coal becomes a greater competitive liability. Yet the focus of governments across Australia is on digging deeper, with the expansion of coal mining the most obvious sign of this. A generation of state and federal bureaucrats, industry lobbyists, political insiders and media commentators have committed themselves to the coal-fired strategy. Rather than forge new competitive advantages for a world where climate change and peak oil are real and imminent threats, they want to cosset the old competitive advantage. Some of the same people who tore down tariff walls to expose uncompetitive industries to free trade are now intent on shielding industries that might be uncompetitive without carbon subsidies. Quarantining the worst emitters and nationalising the cost of their pollution might just hold our precious coal-fired competitiveness together a bit longer. Mining, metals and energy-industry executives running Australian branch offices cling vainly to this hope, but head offices deciding where to locate new production have no attachment to Australia's outdated strategy. They are more dispassionate about investing in facilities with operating lives of fifty years or more. They are looking for a reliable and competitively priced long-term energy supply that is also clean – not what Australia is likely to be offering any time soon.

Internationally, industries are quietly voting with their feet. A competitive long-term clean-energy contract is more bankable than a carbon price carve-out that might disappear at any time. Contrary to what we often hear, the multinationals operating energy-intensive industries in Australia are not moving to countries with no Kyoto Protocol obligations that offer cheap coal with no carbon price. By and large, the shift is to countries offering clean energy. After all, the best way to reduce exposure to rising carbon prices is to avoid exposure altogether by using renewable energy. That's why countries such as Iceland, Russia, Canada, Brazil, Cameroon and the Congo are excelling in the aluminium industry: cheap hydro power is available and reliable. It's why Greenland is suddenly a player in the aluminium industry; it's one of the reasons China built the Three Gorges Dam and why Chinese aluminium producers are eyeing off large new hydro projects in Tibet. They want to use renewable energy to attract the very industries we are led to believe care only about cheap energy. The same trend is apparent in other energy-intensive sectors, such as silicon, zinc and lead smelting. Even in the iron and steel industry, gas is becoming attractive as an energy source, and bio-char is now seen as a potential substitute for coking coal.

Some of the same companies threatening to flee Australia unless they are granted carbon subsidies are also investing billions of dollars elsewhere in renewable energy. Aluminium is the most glaring example. Rio Tinto smelts aluminium with hydro power in Canada, Scotland, Cameroon, New Zealand and Brazil; BHP Billiton does the same in Brazil and is said to be considering a new smelter in the Congo. Rusal and Hydro Aluminium are two other producers who rely heavily on renewables overseas, but not here. Alcoa happily smelts aluminium in Iceland, Canada, Brazil and Ghana with hydro power and is looking to expand with geothermal power. Its executive vice president recently told a European audience: "We are drilling towards the future. Geothermal energy is exactly what the world needs to tap into almost limitless, clean, natural energy and to substantially reduce greenhouse emissions … The technology we hope to develop in Iceland should be applicable wherever there is high-temperature geothermal potential." In Australia, meanwhile, where there is also very high geothermal potential, executives from the same companies talk this way about "clean coal" rather than renewable energy. The rhetoric is completely at odds with the international reality. Our aluminium industry wears green shorts when playing away, black shorts at home.

Yet maintaining the facade of coal-fired competitiveness is a lose-lose proposition. If we hang out the same old shingle as the developed world's polluter haven, and succeed, we will become an even bigger greenhouse ghetto and our carbon liabilities will mount further, both at home and in our exports. If the strategy fails and footloose capital goes elsewhere, Australia is stuck with an inadequate response to climate change, economic losses to show for it, and no alternative strategy. We have little choice but to rethink. Until now, and particularly while the Bush administration pushed a similar line, Australia fuelled global procrastination, knowing it extended the life of our coal-fired strategy. Some may see hope for further procrastination in the fact that Barack Obama is a longstanding supporter of the coal industry, who says "clean coal" can make America energy-independent. Even so, he's taking America in a different direction to his predecessor, and delay will become harder for Australia.

If we are to play our part in an effective global response, we will almost certainly need to cut emissions in Australia by at least 60 per cent by mid-century. In relatively short order, we must stop burning coal unless the emissions are prevented from entering the atmosphere. Peak oil makes the task even more daunting. On top of existing electricity demand, much of our transport energy will have to come from electricity, as petrol and diesel become unaffordable. The challenge will be to green the electricity grid with renewable energy in order to run vehicles cleanly.

Thus, Australia cannot afford the luxury of taking it "slow and steady" or waiting for "clean coal." Like all countries, we must find new opportunities and competitive edges in a carbon-constrained market. Some of what we consider our speciality mightn't survive, but this is a challenge facing many nations as the world adjusts.

Right now, Australia has the opportunity to be an important player in the renewable-energy transition likely to dominate the first half of this century. If we keep delaying, we will be a reluctant passenger. We must take the mental leap and realise that fossil energy is not the only – nor is it the likely – path to prosperity or energy security. That means fostering new competitive strengths. If we can compete cleanly in an energy-intensive industry, well and good. If we can't compete cleanly, then we should get out. If there are Australians who genuinely think that trebling GDP and doubling incomes cleanly are not worth waiting a couple years for, let them say so publicly. For whether they intend to or not, they will help to make the case for change, for dumping our coal-fired game plan.

A generation ago, our leaders showed courage and vision in pushing for unilateral trade liberalisation – they knew it was good for Australia no matter how fast others acted. They were right to turn Australia's economy outward, and the establishment they challenged was wrong. Today the generation that was right on trade liberalisation has much of it wrong on climate change. They now wear the establishment mantle, and it is their turn to be challenged.

THE CARBON LOBBY

Long before climate change and peak oil emerged as serious issues, fossil-energy producers and their biggest Australian customers – our carbon lobby in waiting – exerted a disproportionate political influence. As mostly foreign-owned, capital-intensive businesses, they often lacked electoral clout, so the only real ace in their pack was to threaten to take operations elsewhere. The best way to avoid regulation or to attract government support was to confuse the national interest with their own. That required strong relationships with both sides of politics, with bureaucracies, think-tanks, industry associations, scientific and economic agencies, and media commentators. The extent to which these industries succeeded over a period of decades is best reflected, as we've seen, in successive Australian governments' judgment that cheap coal is the centrepiece of national competitiveness. Thus, when climate change emerged as an issue, there was already a consensus among Australia's business, political and media establishment that the quarry was sacrosanct, coal non-negotiable. It had to be protected at all costs.

In the early 1990s, the Australian carbon lobby got busy, fast. Its members immediately recognised that any Australian commitment to absolute cuts in greenhouse-gas emissions was a problem. It threatened the place of cheap coal in Australian energy and trade policy, raised operating costs and could expose the fact that many commodities were produced with more emissions here than virtually anywhere else. These industries were under threat of losing their privileged position in Australia's political order, but they were well placed to defend it and they knew the issue would take decades to play out. They prepared for a long, drawn-out fight.

Their strategy was to prevent action by Australia, and if that failed, to delay action, and if delay failed, to shift the burden of emission cuts elsewhere. This meant nurturing seeds of doubt about whether climate change was caused by burning fossil fuels. It meant persuading government that emission-intensive industries made a much greater economic and employment contribution than was the case; that greenhouse constraints would wreck the entire Australian economy. It also meant arguing that Australia was a special case: an emission-intensive country with relatively less scope to decarbonise. Finally, Australian action had to be made conditional on action by other countries (there would always be sufficient recalcitrance elsewhere to justify delay). For Australia's emerging carbon lobby, success depended on the political leadership hearing these messages repeatedly from all the sources they trusted. Much of the infrastructure of influence was in place, but the key to winning the policy battle was using big money and the right people. Directly and indirectly, the key players wrote big cheques to the main sources of economic, scientific, ideological, industry, union and political advice for Australian governments.

Knowing that the Australian Bureau of Agricultural and Resource Economics (ABARE) would be relied upon from the mid-1990s as the principal internal source of greenhouse economic advice, a who's who of fossil-fuel producers, burners and users bought chairs on an ABARE steering committee (literally bought: the price was $50,000 per year, and payers included the Australian Coal Association, the Australian Aluminium Council, BHP, CRA, the Business Council of Australia, the Electricity Supply Association of Australia, Exxon Corporation, Mobil Australia and Texaco). This committee oversaw the creation of the economic models on which crucial assessments about emission cuts were based. Though the ensuing analysis showed how easily affordable such cuts were, the presentation was consistently spun to create the opposite impression. Given that ABARE's mission was to "enhance the competitiveness of Australia's agricultural and resource industries" (rather than the broader national interest), the quarry-friendly take on climate change was unsurprising. However, the carbon lobby took no chances, spending large sums on commissioning extra ABARE greenhouse policy work (hundreds of thousands of dollars were spent in one documented case involving the Minerals Council, the Aluminium Council and the Electricity Supply Association of Australia). As a senior carbon lobbyist involved in that work told me:

> Basically ABARE has a requirement to meet certain earnings targets so you can do that through outside consulting. So we commissioned [another party] to do some work … and they got the modelling done by [another party] and ABARE, alright? To our assumptions.

Carbon-lobby companies and industry associations have also contributed considerable sums towards ABARE's overall research program, helping the organisation to raise the tens of millions of dollars it needs annually to meet its external funding requirement. Predictably, all of this has resulted in a steady stream of reports about the cost of cutting emissions that have lent themselves to misrepresentation. A senior minister in the Keating government (still serving today) told me a few years ago that, although he didn't appreciate it back then, ABARE was:

> effectively used by pro-industry interests at the time … If ever you wanted to see a self-serving group, ABARE would be the absolute classic [example] of an agency that acted that way in my view when we were in government. [Q: How do you mean?] They were guns for hire … They were a wholly owned subsidiary of the Department of Primary Industries and Energy at the time and they did the bidding of the DPIE. [Q: So the idea that they propose, that they're an independent …] Is unadulterated crap … [Q: The fact that they are required to raise a certain proportion of their operating budgets from outside consulting, do you think that has an influence on their impartiality?] No, because they were never impartial to start with. I mean – don't fall for any suggestion that there's any impartiality about ABARE.

Knowing that the government listened to more than one source of economic advice, the carbon lobby locked in the key consultants, such as ACIL Tasman, to prepare research backing up its arguments in favour of delay. In recent years, the lobby has also relied heavily on a few consultants in the local branch of CRA International, a firm that has been at the forefront of the successful campaign to delay emission cuts in the United States. Favoured CRA consultants have now moved over to a group called Concept Economics, and together it and ACIL have advised almost every major carbon-emitting industry in the country. ACIL (and CRA to a lesser degree) has also been contracted, sometimes without open tender, to run government inquiries and taskforces, and to provide advice on a host of greenhouse-related matters to federal departments from the prime minister's on down. Thus, whether "independent" advice came from inside or outside government, it was subsidised one way or another by carbon-lobby cash and showed a remarkable but unsurprising uniformity.

Steering the scientific advice to government was equally crucial, and once again the cheques came thick and fast. Since the late 1990s it has been dangerous for large multinationals and their industry associations to deny openly the link between greenhouse-gas emissions and climate change. Rather than risk damaging their brands, they fund front groups to challenge the findings of the UN's Intergovernmental Panel on Climate Change. A small global network of "experts," many of them with no relevant climate-science qualifications, have been funded directly and indirectly to ensure a constant stream of submissions to government inquiries, conference speeches, bogus petitions, documentaries and media commentary. This has created the impression of division in the scientific community where almost none exists.

In Australia since the year 2000, the Lavoisier Group has co-ordinated the local fight against the science. Such organisations can freely say what their sponsors dare not: that climate change is "the greatest fraud ever perpetrated on the public," a "scam" and a "web of deceit," that concern about climate change is "green religion," that wind farms are monuments to "pagan gods," and that the "Great Barrier Reef may actually benefit from some global warming." They publish books like *Thank God for Carbon* and go so far as to suggest that those running the IPCC should "be facing criminal charges and the prospect of going to jail." Internationally, the funding links between climate sceptics and resources companies (especially Exxon and the coal industry) have been exposed. Some of the "experts" are veterans from previous industry-backed campaigns to discredit scientific findings about tobacco, asbestos and leaded petrol. Yet there has been little attention paid to the funding of Australia's sceptics, who have delivered testimony to the US Congress, the House of Lords and at countless conferences. They put their hands on their hearts and say they receive no "research" funding from corporations, but what is not disclosed is who funds the globe-trotting greenhouse advocacy that takes up most of their time.

The funding arrangements in Australia are relatively opaque, but the template is the same as overseas. Occasionally we get a glimpse: Western Mining Corporation (WMC), for example, proudly acknowledged its support for the Lavoisier Group in, of all things, a sustainability report. WMC's former CEO Hugh Morgan launched the Lavoisier Group, and senior staff ran it. More often than not, the sceptics' activities are funded by neo-liberal think-tanks, such as the Melbourne-based Institute of Public Affairs, which are in turn funded by emission-intensive companies. The IPA's Energy Forum is dominated by fossil-energy interests, including La Trobe Valley and NSW coal-fired electricity generators. Senior IPA staff running the forum say the IPA rarely takes a position that differs from that of its Energy Forum funders, "otherwise they'd stop funding us."

To help make these views seem more broadly held, the IPA funds sceptic scientists to speak at industry events. For example, in 2004 and 2005, Australian Institute of Energy members were invited to pay $45 a head to hear one of Australia's most prominent sceptics, Bob Carter, speak in Melbourne and Sydney about "whether climate change is unusual, or justifies policies for costly counter-measures." On its website, the AIE said: "We acknowledge with thanks the sponsorship of the Institute of Public Affairs for travel costs in bringing Professor Carter to Melbourne for this event." The IPA also takes full advantage of the Australian Environment Foundation, a front group designed to look like a grass-roots environmental organisation – a tactic known as "astroturfing." It is chaired by Jennifer

Marohasy, who has been contracted to run the IPA's environmental campaigns since 2003. The foundation claims that emissions trading is "the equivalent of closing down all of Australia's manufacturing and half its rural industries ... [or] the equivalent of closing 72 per cent of our current power generation capacity." It doesn't matter that the greenhouse sceptics have been dismissed by the public as crackpots or that they have had little access to senior bureaucrats – so long as they are taken seriously by enough senior politicians, business leaders and media commentators.

Financing the sceptics is only half the story. At the respectable centre of the debate, a different approach is required. Given that fossil-fuel burning is in fact the problem, the case has to be made that fossil fuels are also the solution – that emissions can be safely and economically stored underground through carbon capture and storage. Funds have been channelled into sections of the CSIRO to support research and development of CCS technologies. This has helped to meet external funding requirements just as it helped ABARE, but it has also created "constant pressure" to focus on commercial research for paying industry partners. As CSIRO has publicly acknowledged, it also gave polluter clients the right to veto research findings they didn't like being made public. A succession of eminent scientists have spoken out since leaving the CSIRO about the various ways in which they were barred from making public comment unwelcome to the carbon lobby or a government determined not to offend it. Talk about emissions trading, emissions targets or climate refugees was off limits – as was membership of the Wentworth Group of Concerned Scientists.

Various Cooperative Research Centres, along with the Centre for Low Emission Technology, have been established with carbon-lobby backing. On the surface, they resemble academic institutes objectively studying the viability of CCS. In reality, they are champions of the technology whether it is viable or not, and PR conduits for those providing the funding. While these "R&D institutions" are busy generating the impression that "clean coal" is a fait accompli, their sponsors are busy inflating hopes of what it might achieve. The Hydrogen Energy collaboration between BP and Rio

Tinto, for example, promoted the idea that coal gasification is the path to a hydrogen economy. The Monash Energy collaboration between Shell and Anglo Coal promoted the idea that coal liquefaction can deliver Australia independence from foreign oil.

Though the carbon lobby is bashful about the amount of money it funnels to independent scientists, the Australian coal industry has consistently lauded its own generosity in placing a levy on itself to fund carbon capture and storage. The billion dollars this levy is intended to raise sounds impressive, but it is spread over more than a decade. It is a tiny fraction of the industry's profits, and not enough to cover the cost of just one large conventional coal-fired power station, let alone a "clean" one. The aim hasn't been to clean up coal but to clean up its reputation – to keep the idea of clean coal alive politically: a "permanent alibi" as Robert Manne recently put it. For so long as CCS *might* be "viable" one day, the industry bets that no Australian government will countenance coal's demise. To that end, the coal industry and its proxies push the argument that the technology to save coal might not make it "onto the shelf" if too much money and effort is devoted to getting renewable technologies "off the shelf."

Most major greenhouse polluters are represented by at least one industry association: mining, electricity, cement, oil, car manufacturing, trucking, paper, plastics and chemicals. Some, like the Australian Aluminium Council, do little but greenhouse policy; all spend large sums communicating the economic and scientific cases for delay or taking minimal action. Their willingness to spend has enabled these industries to dominate the debate. With a steady stream of international negotiations to be monitored, and a plethora of domestic policy processes, very few stakeholders can afford to be everywhere. Greenhouse policy is a rich man's game and the deep pockets of the Australian carbon lobby have made its members ubiquitous. From parliamentary testimonies to press-club speeches to the relentless stream of communication with ministerial offices, senior bureaucrats, party officials and the press gallery, they've never missed a day at school.

To make their dollars go further, polluters have channelled much of their lobbying effort through the Australian Industry Greenhouse Network (AIGN), an alliance representing almost all of Australia's biggest fossil-energy producers and consumers, either directly or through their industry associations. In interviews recorded with me on the condition that their names were not revealed, a dozen past and present lobbyists from the AIGN described at great length how they had effectively hijacked Australia's climate-change response. They affectionately referred to themselves as the "greenhouse mafia" and "the mob," names befitting both disproportionate behind-the-scenes influence and the strength of their grip. AIGN membership looked broad enough to give the impression – so desperately sought after by government – that industry was speaking with one voice. This helped to drown out competing voices, as did the carbon lobby's success in paying large sums to join and dominate organisations one might expect to have a more constructive influence on the debate. Using relatively small stakes in cleaner energy to get in the door, companies whose focus is oil or the unfettered expansion of coal mining and burning dominated the Australian Gas Association and, more recently, the Clean Energy Council. By investing relatively greater resources, some of the same companies have set the greenhouse policy direction of umbrella groups like the Business Council of Australia, the Australian Chamber of Commerce and Industry, and the Australian Industry Group.

Both major political parties have received millions since 1998 from organisations that are directly or indirectly represented by the AIGN. On the Liberal side, there is relatively more direct corporate support from carbon-lobby companies and industry associations, but Labor receives plenty and it enjoys a virtual monopoly on money donated by interested unions: the Construction, Forestry, Mining and Energy Union (CFMEU), the Australian Workers' Union (AWU), the Transport Workers' Union of Australia (TWU) and the Australian Manufacturing Workers' Union (AMWU). The most prominent such group is the Mining and Energy Division of the CFMEU, which claims to be "the principal union in the black and brown coal mining industries." In 2007–08, its $700,000 donation was the largest external contribution made to the national secretariat of the Labor Party. In 2006–07, its $300,000 donation was the second-largest. In 2004–05, its $430,000 donation was again the largest.

In politics, advice from trusted faces hits the mark more often, and most of the key lobbyists and industry association heads have worked previously as senior officials or ministerial advisers in the federal resources, energy, trade or transport portfolios. This is also true of the favoured economic consultants. When senior public servants and ministerial advisers take the calls of these people, they are talking to former colleagues, and often ex-bosses. These are relationships where the person calling still has the whip hand.

So, for instance, when Woodside's Liberal-leaning lobbyist called John Howard's environment minister, it was the former chief of staff of that office on the line; when Woodside's Labor-leaning lobbyist rang, it was the former general secretary of the Labor Party. When the LPG industry's lobbyist rang, it might have been the former federal director of the Liberal Party. When an economist acting for the LNG industry calls now to discuss emissions trading, it's probably the former head of ABARE on the line. When Rio Tinto's chief economist calls, it's his former deputy. In all sorts of ways, the person calling on behalf of the carbon lobby is able to pull rank. Once, when the head of the Australian Coal Association called, it was an ex-prime-ministerial adviser on the line; now it's Australia's former ambassador for the environment, as is the case when the Alcoa Foundation rings today. For many years, if BHP Billiton's external-affairs boss called a person in the industry department looking after greenhouse policy, the person receiving the call was not just talking to the former departmental head, but to a former prime-ministerial adviser. Now when the BHP Billiton external-affairs boss rings, it's a former adviser to *two* prime ministers and one premier on the line, not to mention an ex-national secretary of the ALP. Until recently, when Rio Tinto's external-affairs manager called ministerial offices, it was the PM's nephew calling. For a long time, if Rio's chief technologist called the federal government's chief scientist to spruik "clean-coal" technology, he would have been talking to himself!

Board appointments have been another great way to lock in access and intimidate government. Within months of his departure as head of the Department of Foreign Affairs and Trade, the late Ashton Calvert joined the boards of Rio Tinto and Woodside – two companies determined to avoid emission constraints. Another company with a strong interest in keeping coal exports out of the greenhouse policy discussion, the coal-mining contractor Theiss (owned by Leighton Holdings), invited former ALP environment minister Ros Kelly onto its board years ago. Meanwhile, Macarthur Coal, one of the companies with a big interest in digging up Queensland's coal, is chaired by former Queensland treasurer Keith De Lacy. Another former Queensland treasurer, David Hamill, chairs the board of Babcock and Brown Infrastructure (BBI), which has a huge interest in shipping coal out of the same state. BBI's cousin, Babcock and Brown Power, which burns some coal, recently made the former head of the Queensland premier's department, Ross Rolfe, its chief executive. With each of these appointments, the fossil-energy lobby subtly tightens its grip on the country.

Recruiting from the bureaucracy and from both sides of the partisan divide ensures that "trusted face" status is election-proof. If anything, elections serve as opportunities for trusted faces to assume more powerful positions. Gary Gray and Andrew Robb, for example, ran the Labor and Liberal parties respectively in Canberra, and then took carbon-lobby work before being elected to parliament. The election has elevated their industry-friendly (and in the case of Robb, overtly sceptical) views to opposite front benches. The 2007 election also catapulted Jason Clare, long-time manager of corporate relations for Transurban, the biggest owner of Sydney and Melbourne toll roads, into parliament and onto the House of Representatives Committee on Infrastructure and Transport. When carbon-lobby recruits aren't moving through the revolving door between government and industry, they're often moving sideways between industry associations in a game of musical chairs (I like to imagine the theme from *The Godfather* playing in the background). Aside from providing a career path for lobbyists, this engenders solidarity and conformity. When a CEO is replaced, it's by someone with the same world view: that fossil fuels are sacred.

Having recruited trusted faces, the big emitting sectors have ensured that they appear in all the right places, amplifying the delay case from all sides. It's no accident that the Institute of Public Affairs has been dominated by mining company executives, or that eleven of the thirteen presidents of the Business Council of Australia have been directors or executives from the resources sector. It's no accident, either, that the carbon lobby has dominated almost every greenhouse-related consultative committee established by the federal government and its agencies.

The carbon lobby tries half-heartedly to downplay its influence. Though run by registered lobbyists, the AIGN shamelessly claims not to be a lobbying outfit at all. Its leaders dismiss the image of "balding, cigar-chomping" men "in some dark smoke-filled room chortling about Cabinet submissions they had written and policies they had manipulated." "Ruthless!? Good heavens, no, we're angels. Simply trying to make a dollar!" one laughed during an interview with me. Anyone who works in greenhouse policy circles knows better: with almost all the trusted sources of advice to government, the fix is in. For some, this is the key to advancement; others grin and bear it; most shake their heads. Outwardly the debate looks lopsided, but different views are expressed and environmentalists get their say. Government seems to take different opinions into account. However, any appearance of pluralism is deceptive. At senior bureaucratic and political levels, the same few messages are being heard from the same few sources. Policy is contaminated by patronage at every turn.

The internal economic advice might come from ABARE, but much of the underlying money and oversight comes from the carbon lobby; the external economic advice might come from an "independent" consultant, but carbon-lobby cash and assumptions are usually behind it; the scientific advice might come from CSIRO, Cooperative Research Centres or the chief scientist, but there is still carbon-lobby cash, oversight and censorship; behind the think-tank agit-prop, more polluter cash and oversight; and more again behind political fundraising. Policy advice might come from federal departments and have the same authors as polluter advice to government. As I documented in my book, *High & Dry*, former bureaucrats working as carbon lobbyists have been invited into government departments to draft Cabinet submissions and ministerial briefings. It has changed this year, but for a long time when Australia's delegation went to climate negotiations, the minister found senior government officials on one side and the carbon lobby on the other – as part of the official team. If the minister got confused about who was who, it mattered little because, if history was any guide, the official would soon be the lobbyist in any case.

Sometimes the carbon lobby looks lucky: when the AIGN boss was seconded into the prime minister's department to help design an emissions trading scheme; when the carbon lobby was secretly invited to develop government policies, funding and PR for "super dooper" low-emission technologies (and the renewables sector cut out); when John Howard consulted a who's who of big emitters before rolling his Cabinet to stop emissions trading; when a sceptic was made head of the Cooperative Research Centres Association and when another was put in charge of handing out federal money for low-emissions projects. Sometimes lightning even strikes twice: as when a former coal-industry PR man was made the *Australian*'s environment reporter in the run-up to the 2007 election, then made CEO of the Clean Energy Council – the perfect place to emblazon "clean coal" with the clean energy brand. In reality, however, the carbon lobby makes most of its own luck. It's only thanks to years of hard work by those with the most to gain from delay that the wolves run the henhouse.

Many people wonder how the wolves sleep at night, how they justify their actions to themselves and to their children. Some compare individuals prominent in the denial and delay campaign with war criminals and call for them to be put on trial and ultimately incarcerated for crimes against future generations and the environment. They imagine genuinely evil people knowingly fuelling environmental catastrophe to feather their own nests. Yet in the vast majority of cases, this is wholly to mischaracterise the people working for our worst polluting industries. Almost all sleep well at night because they do not believe what they do is evil; on the contrary, they believe they are on the side of light. For most of the neo-liberal think-tanks and front groups that promote the climate sceptics, and for their acolyte media columnists, anything that hinders economic freedom is anathema. Suspect at best, global warming is just another trumped-up attack on the free market. They are serious when they dismiss climate change as a "new upper-class religion," when they call Al Gore's *An Inconvenient Truth* "bullshit from beginning to end."

As for the professionals – the industry association bosses, the lobbyists, spin doctors, the "clean-coal" scientists, the economic modellers and even the sceptic scientists – some are indeed in it for the filthy lucre. The carbon lobby has rescued plenty of mediocre and obscure careers. However, many are genuinely proud of their actions in the greenhouse debate. They believe emission cuts by Australia *are* largely futile, that carbon will leak offshore in great quantities, that clean coal can save the day. Some scientists promoting CCS do think they are part of the solution; some economists are genuinely troubled by shaving even the tiniest slice off GDP growth. I have no doubt that some sceptics are sincere in their doubts. Where there is overlap between the neo-liberal zealots and the professionals, there is even less sense of shame. There is the usual strong tendency towards "role morality" ("where I stand depends on where I sit"). The dot-points are familiar, comfortable and advanced with conviction and pride. They know they run the henhouse, but not that they are the wolves.

Not seeing them as wolves, John Howard left the henhouse door wide open. Using the right people made the carbon lobby more effective, but the right prime minister made them invincible. Here was a leader who said Australia should never even have got involved with the global climate-change treaty forged in Rio in 1992. The carbon capture of Howard's government was as easy as it was thorough. Howard's grip was autocratic to the extent that delaying emission cuts merely meant convincing him. The sceptics had direct access to the prime minister and real influence on his government. As one senior Lavoisier figure told me in a taped interview, thanks to his organisation "there's an understanding in the Cabinet that the science is all crap." Howard was surrounded by sceptics, including his party president, some party fundraisers, some senior ministers, the chair of his environment policy committee, not to mention many on the backbench. With Howard having taken charge of greenhouse policy and repeatedly intervened to protect the interests of our worst polluters, there seemed no prospect of an effective response to climate change. Without him, there seemed at least a chance of forward movement.

However, Labor's stake in the status quo was larger than most realised. Various AIGN lobbyists and prominent sceptics had spent their formative years serving the Hawke and Keating governments. They had helped to devise Australia's coal-fired export strategy and the reforms accompanying it. Turning the Australian economy outward to make it more trade-oriented was something they viewed as the grand contribution of their age. Shared responsibility for this economic legacy bonded the carbon lobby with federal Labor while it was in Opposition and with Labor state and territory governments around the country.

The coal-fired strategy was also familiar to Kevin Rudd: he had worked for the Department of Foreign Affairs and Trade during its ascendancy; and as chief of staff to the Queensland premier, he'd been a central figure in the most coal-friendly state government in the country before his election to federal parliament. Some of his former associates now sat on coal-company boards. Having persuaded Labor's national conference to dump the three-mines uranium policy, Rudd made Sir Rod Eddington a key business adviser on greenhouse policy. Eddington sat on the board of Rio Tinto and was far from the only voice in Rudd's ear that was as close to the carbon lobby as those Howard had trusted. During the 2007 election campaign, Rudd gave the impression that he would put the national interest ahead of polluter interests, that Australia would be more than a quarry, and that he and John Howard would be like chalk and cheese on climate change. Time would tell.

Labor was not elected on a greenhouse policy platform so much as a political pedestal. After eleven years of obfuscation, the priority for many people was removing John Howard – not Kevin Rudd's policy offering. Out-greening his rival in the 2007 election campaign was never going to be hard. The Coalition recognised climate change was a voting issue, but its response was too little too late. Having claimed for years to "lead the world on climate change," Howard found it difficult to persuade a spin-weary public that he had finally come round. He tried announcing his way out of trouble, but it fell flat. It was hard to take the Sydney Declaration at APEC in September 2007 seriously when it committed no country to emission cuts, ever. Few bought Howard's implied support for a new emissions target; he wouldn't nominate any figure and, after all, this was the same man who had argued for years that "targets and timetables" didn't work, that "new technology" was the real answer. And when Howard pledged to replace the renewable-energy target he'd consistently refused to increase with a new "clean-energy target" for which nuclear and "clean coal" would be eligible, it failed the sniff test altogether. According to John Hewson, when Malcolm Turnbull was asked as environment minister how he would deliver on the promise to phase out incandescent light bulbs, his reply was, "Fucked if I know!" It was last-minute, desperate stuff, and every man for himself, as the leaked Turnbull plea to the Howard Cabinet to ratify Kyoto only reinforced.

Labor, not surprisingly, got the benefit of the doubt. Kevin Rudd provided sufficient hope, ticking off many of the commitments Howard had avoided for so long: Kyoto ratification; a 60 per cent cut in emissions by 2050; an emissions trading scheme commensurate with that target starting in 2010; a 20 per cent renewable-energy mandate; and, finally, a full assessment of the economic impacts of climate change and emission cuts on Australia, to be undertaken by the prominent economist Ross Garnaut. All this had the desired political effect: Rudd was swept into office with the help of climate-change voters. What many either failed to notice or dared not say was that Labor's policy platform was no guarantee of an effective response to climate change. Most of the big decisions had been deliberately postponed, and Labor had left itself lots of wriggle room. The hope was that Rudd wouldn't use it.

The political equation facing Labor was unenviable. Granted, the public acknowledged the seriousness of the climate crisis and demanded serious action. On the other hand, Australians repeatedly tell pollsters they're happy to pay more for clean electricity, yet only a tiny fraction have signed up for green power. Politicians fear that voter behaviour might mirror this consumer behaviour at future elections. Polls also suggest that the public wants government policies that lead to deep cuts in Australia's greenhouse pollution, even if this means paying more. Yet governments suspect only very few of us really mean it, especially if higher petrol prices are involved. Having promised deep emission cuts, the new government was spooked and tempted to find a way to defy logic. The deeper the emission cuts in Australia, the higher the carbon price; the more businesses and people shielded from a carbon price, the less it cuts emissions; the more people are compensated, the greater the carbon-cost burden for others. The only way around these realities would be to conjure up the illusion of an effective response, in which no one felt worse off. In flicking the hard decisions to Ross Garnaut's climate-policy review, Labor bought time to decide between politically risky action or the illusion thereof. Most of the political atmospherics favoured the latter course.

It was as if the Australian economy's "best before" date coincided with Labor's election. As 2008 wore on, the global credit crunch, the prospect of a deep worldwide recession, and a commodity-price crash knocked the wind out of the economy. The stock market and the dollar went into free-fall. Australian consumers retreated to their devalued houses to ponder the remnants of their superannuation, their frozen managed funds, and whether their bank savings and jobs were safe. With business and consumer confidence at rock bottom, greenhouse policy became even more politically perilous. Internationally, economic catastrophe had quickly and predictably supplanted climate change and the so-called War on Terror. When times were good, the carbon lobby had argued that it was madness to mess with a winning formula; now times were very bad, they could instead say it was "not a time to experiment with the Australian economy." In truth, deep emission cuts would never suit our worst polluters, but with Australia likely to negotiate a new emissions target in late 2009, an emissions trading scheme due to start in 2010, and a federal election soon thereafter, the last thing the new government needed was a perception that green zealotry might tip Australia into recession.

In Canberra, most of the greenhouse-policy furniture wasn't picked up by removalists when the Lodge was vacated – it was too firmly bolted to the floor. Many of the senior bureaucrats who had ardently backed Howard's greenhouse response (not merely implemented it) remained in place. Some shifted from the Department of Foreign Affairs and Trade to the new Department of Climate Change; some took jobs with carbon-lobby industry associations and economic consultancies, further bolstering the greenhouse mafia's position. The economic and scientific agencies that advise government continued to depend on fossil-energy patronage to meet their external fundraising requirements. Treasury and the newly created Department of Climate Change would assume a greater responsibility, but installing a few new structures and faces didn't change much. At the decision-making centre, Australia's most effective and notorious "iron triangle" of polluter, political and bureaucratic elites survived. The policy consensus among senior officials, the carbon lobby and the elected government moved, but not nearly as far as many imagined.

The major shift was to extinguish the possibility that Australia might avoid *any* post-2012 emission constraint – while Howard survived, some Canberra elites still nurtured that dream. On balance, Labor's 2050 target was seen as desirable: it provided certainty about the long-term direction without mattering much today. Kyoto ratification was good for Australia's international reputation but otherwise inconsequential – like backing the winning side in a blow-out match seconds from full-time. And a polluter-friendly emissions trading plan (already foreshadowed by Labor) was something those inside the iron triangle had been working up for years. The incoming government's persistence with an expanded renewables target of 20 per cent was inconvenient, but it might not last. There were also signs that Labor might eventually come around to nuclear power: the former NSW premier Bob Carr backed the idea, and the head of the AWU would soon be arguing that "if we are going to be a green Labor government, then we have to look at nuclear." What mattered in the short term to the mandarins was retaining Labor's pro-"clean-coal" position; making Australian action conditional on that of other countries; minimising the target Australia would accept in Copenhagen in 2009; and using every imaginable tool to reduce the cost of meeting it, especially for the most emission-intensive sectors. The real challenge wouldn't be accommodating a radically different government policy; it would be helping the government make a potentially very accommodating policy look like a radical departure.

The presence of a small but vocal group in Labor's own ranks that agreed or sympathised with the sceptics would make this task harder. The Lavoisier Group still had good contacts in high places: its president was a former Labor finance minister, Peter Walsh. As a senior Lavoisier office-bearer told me: "We have a good following in the Labor Party ... Walshy has been a fantastic president and he's given us entrée to the Labor Party because he's still got a fan club in the Labor Party and quite an effective one too ... Walshy has access to anybody he wants to see apart from the Labor Left. If Walshy wants to talk to the prime minister, in the end he'll get to see the prime minister."

Sceptics could expect a sympathetic ear from Martin Ferguson, the new energy and resources minister, and perhaps also from Gary Gray, a former Woodside lobbyist and now parliamentary secretary to the minister for transport and infrastructure. They could also rely on the support of AWU patriarch Bill Ludwig, the biggest Labor powerbroker in the prime minister's own state. At the Parliament House launch of the Lavoisier Group's *Nine Facts about Climate Change* in early 2007, Ferguson reportedly told an appreciative audience that "Labor [is] going to wait for clean-coal technology to come along before doing anything to the coal industry."

The post-election political contest favoured inertia. In July 2008, the Coalition parties lost the balance of power in the Senate, but the government did not obtain a majority. From now on, it was simpler for Labor to do business with the Coalition to pass legislation rather than rely on unlikely alliances among Green, Family First and independent senators. It made political sense to chart a middle course rather than try to outgreen the Greens – to "get the balance right" between the sceptics on one hand and "radical green groups saying that we haven't gone far enough because we haven't closed down the coal industry by next Thursday," as Rudd put it.

Meanwhile, on the Coalition side, greenhouse policy hadn't really shifted since John Howard's departure. With Kyoto ratified and nuclear power off the official agenda, two prominent Howard positions were abandoned without any consequence. Recycling the Liberals' pre-1996 rhetoric, and perhaps inadvertently conceding failure on Howard's part, Brendan Nelson said Australia must stop living off environmental capital. But the people he chose for his shadow front bench ensured that Nelson's opposition would be as dominated by greenhouse denial and delay as Howard's government had been. In the Senate, the Liberal Party leader Nick Minchin (*sceptics should be applauded for speaking out*) and deputy Eric Abetz (*weeds are more serious than climate change*) remained fervent sceptics. Fellow sceptic Ian Macfarlane (*Gore's film is incorrect "entertainment"*) kept the industry portfolio. After weeks of cringe-worthy internal division over the party's thin green coat of paint, Nelson conceded that his climate-change policies were basically the same as those of Howard. Nelson's successor would later acknowledge this too.

When Nelson was ultimately despatched, some (me included) hoped Malcolm Turnbull might set a new course. Having reportedly stood up to

Howard in Cabinet over Kyoto ratification, he might have given Kevin Rudd competition on climate change – much as the UK Conservative Party has done under David Cameron, and as Arnold Schwarzenegger has in California. It wasn't to be. Turnbull left sceptics in key portfolios: Macfarlane got energy and resources, and the new emissions-trading portfolio went to Andrew Robb (*the science is unproven ... most of the warming is natural*), a man who had not merely sided with sceptics but who had dismissed climate change as a cause célèbre seized on by lefties with nothing to do since the collapse of communism. Turnbull appointed a sceptic as chief of staff and Henry Ergas, the chairman of Concept Economics, was commissioned to advise on how to spend emissions trading revenue. Ergas had earlier suggested in the *Australian* that it was more efficient to let the planet warm than to cut emissions. The National Party, meanwhile, elected Barnaby Joyce (*ignore these environmental goose-steppers*) as its Senate leader – he was more interested in Australia leading the world in Antarctic coal mining than in cutting greenhouse emissions. Not surprisingly, when Turnbull finally announced his climate-change policy – without any meaningful details – its common thread was the avoidance of anything that might price carbon or constrain industrial emissions.

Had the Coalition created a real contest on greenhouse policy, had the Greens gained serious momentum, had the economy not turned sour when it did, and had the iron triangle in Canberra been looser, Kevin Rudd might have done much more of his political gardening on the green side of the debate. Instead, all the pressures of the day inclined the canny politician toward the appearance of action over the reality.

Illusion would also do just fine for the burgeoning ranks of the "greenshoe brigade," who could make plenty of money whether or not Australia's emissions actually fell. To those eyeing a new income stream from emissions trading or renewable-energy mandates – the banks, law firms, accountants, brokers, carbon off-setters, even renewable-energy companies – what mattered was the creation of a new market with all the necessary accessories. It didn't *have* to be one that led to deep cuts in emissions.

Most green-lifestyle entrepreneurs looking to cash in on public concern about climate change didn't even need a carbon price to market everything from beer to funerals as climate-friendly. In casting their vote for the new government, many people had done what little they felt they could to influence the big picture, and their attention now understandably turned to minimising their individual greenhouse emissions. It was something they could control, something in which they could take satisfaction.

In tandem with the markets, the media seized on this impulse. ABC TV screened *Carbon Cops* and almost every magazine rushed out with a "green issue." Greenhouse narcissism and climate-change news merged. For almost a year the *Sydney Morning Herald* devoted a section of its website to the "Earth Hour" it sponsored, presented in a way that made it indistinguishable from real environmental news. Even the *Australian* jumped on the bandwagon, announcing "Green Awards" for companies and individuals fighting global warming, despite having in its editorials only recently dismissed concern about emissions as "a new climate-change religion." In the long run, individual action, and the burgeoning number of green businesses and media catering to it, is undoubtedly positive. It does not cut our overall emissions significantly, but over the long term it gradually shifts the boundaries of acceptability – for individuals, communities, businesses and government. Unfortunately, in the short term, the busier we are kept doing our bit, the easier it is for government to avoid doing *its* bit to ensure that Australia's emissions fall.

For Australia's carbon lobbyists, life without Howard was a new habitat, but one to which they easily adapted. The global financial crisis gave their cause a political tailwind, and their base was broader than ever before. The AIGN's ranks now included the Australian Trucking Association, the Australian Food and Grocery Council, the National Association of Forestry Industries and the Australian Industry Group. They could rely on the increasing political clout of – and vocal public support from – sympathetic unions, particularly the AWU and the CFMEU. They also had strong backing from Labor governments in Queensland, New South Wales, Victoria and Western Australia, all of which were frantically accommodating greater coal and LNG production.

Having renewed itself and its networks to suit the Labor scene, the AIGN and its members bombarded the political process as never before with every imaginable argument to minimise emissions targets, delay emissions trading and lock in exemptions for themselves. Leaders of the coal, aluminium, petroleum, LNG, cement, steel and trucking industries (and their unions) took to the stage. Woodside's Don Voelte, the AWU's Paul Howes, Australian Industry Group's Heather Ridout, the Minerals Council's Mitch Hooke, and the rest of the AIGN doomsday squad warned of economic apocalypse if Australia cut emissions too soon. It was reminiscent of the memorable scene in Monty Python's *Life of Brian* where prophets of the false, boring and "blood-and-thunder" varieties compete for would-be disciples. Their faces painted, stomping the floor, they warn of demons and blood, of three-headed serpents and nine-bladed swords, desperately trying to top one another with ever more outlandish fire-and-brimstone predictions. At least in *Life of Brian*, onlookers were suspicious: "He's making it up as he goes along!" says John Cleese. Our polluters faced a more gullible audience.

Electricity generators started by saying that emissions trading could plunge Australia into darkness. Using "independent" research from ACIL

Tasman, the Energy Supply Association of Australia said emissions trading might slash the value of coal-fired power stations and cut profitability so deeply that stations in at least four states would have to be needlessly mothballed. The wholesale price of electricity would rise from 35 to 55 per cent in every state but Tasmania. The National Generators Forum warned of blackouts, and, using "independent" research from CRA International, claimed that gas prices might double along the eastern seaboard. The oil and gas industry used the same consultants to come up with a 24 per cent electricity-price increase. As it happens, real incomes are also projected to grow rapidly, so even if the carbon lobby's hyperbole is half right, people will spend a similar proportion of their income on energy over the next few decades as today even if Australia halves emissions.

The generators flagged their intention to bill the federal Treasury for their own stupidity. For twenty years, carbon pricing had been anticipated by business. Back in Kyoto in 1997, even the Howard government had made emissions trading a precondition of Australia signing a global agreement. Yet generators somehow missed the memo and kept writing cheap, long-term electricity deals with Australia's most emission-intensive businesses – even after Kyoto. Now, in 2008, they said no one could have seen this coming: emissions trading was an "arbitrary destruction of shareholder wealth through a dramatic change in government policy" that "could not have been reasonably anticipated." To avoid a catastrophic new Dark Age, government must "compensate" generators in full for the inevitable drop in the value of their assets. One wonders how the generators argued this with a straight face, but they managed.

Other large greenhouse polluters warned of job losses and billions of dollars in lost investment. The aluminium industry, the steel, lead and zinc industries, cement, LNG and paper would all pack up and move offshore unless they were carved out of emissions trading or fully compensated for its impact. The Business Council of Australia claimed that half of Australia's emissions-intensive, trade-exposed firms would either close or lose one- to two-thirds of their pre-tax earnings and be forced to "fundamentally review their operations." All such firms had to be fully compensated until their international competitors faced an equivalent carbon cost. "We don't want to be alarmist about it ..." said the Australian Industry Group, but all up a million Australian jobs were at risk. Rather than having a real emissions trading scheme where polluters paid a meaningful carbon price, the industry body recommended a "dry run," so that polluters could get used to the scheme and go through the motions, without actually paying anything.

Polluters and their unions sought a blanket exemption for facilities not yet even built. This would be a "massive carrot" for new energy-intensive investment, according to the AWU's Paul Howes. Such projects would be "world's best environmental practice," he argued, apparently unaware that world's best practice means zero emissions and renewable energy, not what would be used in Australia. Without such a carve-out, it was "a certainty" that the industries that "pay our bills" would leak offshore, said Howes.

In reality, very little of Australia's economy is much exposed to carbon pricing, and the number of genuinely vulnerable jobs is a small fraction of the million claimed by the Australian Industry Group. The deceptive impression given by the blood-and-thunder prophets was that entire industries would shrink, with commensurate job losses. When measured against a fantasy-land scenario in which coal would be burned for decades with impunity, they claimed that there would be huge percentage falls in output across a range of industries. What they concealed, and what Treasury's modelling would later show, is that almost every Australian industry – including almost every quarry industry – would continue growing, only more cleanly, and emerge far bigger in 2050 than today.

As for carbon leakage, the chance of this happening on any significant scale is virtually nil. As John Hewson once memorably told me, "You just don't throw an aluminium smelter in a backpack and take it off to Indonesia." In almost every case, having sunk so much capital in their Australian operations, companies would instead wear a slightly lower rate of return in the short term and, focusing on the long term, they'd look damned hard for efficiency gains. Multinationals might focus *new* production in countries where low-emission energy was abundant and competitively priced, but they wouldn't do this purely on the basis of carbon price. Any company executive who did would be a laughing stock, for they would be betting against all odds that the carbon price in the new location would not change for the next four to five decades. They would also be ignoring the myriad factors that have a far greater impact on profits and competitiveness: skills base, political stability, tax rates, infrastructure, reliability of energy supply, and exchange rates. Most large resource multinationals deliberately have facilities in numerous locations, reducing their exposure to all these risks through diversification. Sure, they routinely threaten to relocate production if government doesn't give over a large wad of cash or tax break for this or that. It's standard operating procedure for which an army of spinners is used. (For a time, I confess I used to deliver none-too-subtle hints for BHP Billiton in this vein to the Howard government and its bureaucrats.) But you don't uproot all your production to chase a better carbon-price deal – it just doesn't work that way.

But let's imagine for a moment that it did. What if Woodside followed through on its threat to wind back operations on the North West Shelf and float its processing facility out of Australian waters into East Timor just to avoid a carbon price? What if the aluminium industry, Sun Metals, Nyrstar and others actually meant what they said? What if these "carbon pirates" were ready to sail the high seas in search of the lowest carbon price, and preferably none? Such an approach would be reprehensible in several ways. It would knowingly add millions of tonnes of extra greenhouse pollution annually to protect a small slice of margin; it assumes that developing nations dare not apply a carbon price comparable to Australia's for decades; and it implies that such companies would lobby the world's poorest countries to ensure that a carbon-price-free environment stayed that way. It forfeits any claim a company might make to good corporate citizenship.

Of course, in the event that carbon pirates did flee, the emissions picture for Australia would improve. If those sectors seeking compensation were put in a backpack and transported offshore tomorrow, Department of Climate Change figures suggest that around 94 per cent of GDP might survive, while Australia's emissions would drop by more than 40 per cent. In theory, we'd be halfway towards the deep cuts targeted by 2050 in one fell swoop and well on the way to a cleaner economy. In practice, they will not flee. We needn't wait for emissions trading to confirm this, either. As we've seen, companies predicting carbon leakage from Australia already produce the same commodities offshore with renewable energy and vastly lower emissions-intensity. The best-known aluminium producers (Rio Tinto Alcan, Alcoa, Hydro Aluminium and Rusal) all smelt the vast majority of their aluminium with renewable power. As for LNG, industry analysts believe Australian projects will not be rendered uncompetitive if producers are forced to pay for their emissions.

Not surprisingly, ABARE found in 2007 that even if Australia acted alone in cutting greenhouse emissions, for every tonne of greenhouse gas cut only around one-tenth of a tonne would leak offshore – and compensating polluters with free permits made little difference to that. Treasury's 2008 modelling drew the same conclusion, saying: "There is little evidence of carbon leakage ... fears of carbon leakage may be overplayed." It also found that shielding industries, even the aluminium industry, made no difference to their output in Australia in the long run. In spite of this, the blood-and-thunder prophets showed just how easy it was to distort the policy debate with unsubstantiated threats and self-serving economic analysis. The big question for much of 2008 was whether Ross Garnaut would see through it all. Hopes were initially high.

ENTER ROSS GARNAUT

The stage was set for Ross Garnaut to emulate Sir Nicholas Stern's groundbreaking work for the UK government. At first, people liked what they heard. The contrast with the Howard years couldn't have been more stark. Garnaut provided the clearest clarion call yet about the seriousness and urgency of the environmental threat facing Australia. He didn't mince words: environmental treasures like the Barrier Reef and irrigation-based agriculture in the Murray-Darling were very likely doomed. Gone was the interminable reluctance to contemplate emission cuts. Garnaut not only assumed Australia would make deep cuts, he said they should be made sooner rather than later. Polite, yet resolute, he demolished the monotonous excuses for delay. Gone was the spin about Australia "already leading the world" on climate change. Garnaut opposed giving generators free permits, and he said motorists shouldn't be shielded from the impact of emissions trading on petrol prices. To many, it seemed as if the Elliot Ness of climate change had finally arrived to take on the greenhouse mafia. The tone was different from anything emanating from Australian governments in the past decade. It was as if reams of talking points had been shredded.

As it turned out, not all of them were, and Garnaut was no Elliot Ness. Having raised hopes by his public comments, Garnaut quietly and comprehensively dashed them in the fine print of his reports. He recommended that emissions trading begin in 2010 to help deliver Australia's Kyoto target, but five years' notice should be given before moving to any new emissions trajectory. Even if Australia committed to a new target in Copenhagen in 2009, Garnaut was in effect saying Australia should not deviate from John Howard's proposed path until around 2015. If it did, that should be conditional on the action of other countries (pretty much Howard's line too). In his final report, Garnaut said Australia should offer to cut emissions by 25 per cent by 2020 as part of a comprehensive global agreement targeting 450ppm CO_2-e.* Calling this unlikely, however, he

* The combined level of greenhouse gases in the atmosphere is measured as CO_2-e parts per million. The present level is 455ppm, which includes 386ppm of CO_2.

said Australia should only offer to cut emissions by 10 per cent by 2020 if a global agreement targeting 550ppm was achieved. If no agreement at all were to be reached in Copenhagen, he recommended 5 per cent emission cuts by 2020 as an unconditional offer. The 10 per cent option was likely to be "the best available to us now."

Amazingly, no one questioned the way in which Garnaut deemed the 25 per cent target "unachievable" on the basis that it was "consistent" with a global agreement targeting 450ppm, whereas his 10 per cent target was "achievable" because it was "consistent" with the more realistic 550ppm global goal. On the contrary, some scientists and environmental campaigners embraced this logic, saying that Garnaut's 5–10 per cent targets for Australia "locked in" the destruction of the Great Barrier Reef and various other catastrophic consequences. Both Garnaut and his critics disingenuously linked hypothetical Australian emissions targets and global parts per million scenarios. Behind the link was the idea that national targets should reflect an eventual "contraction and convergence" of per capita emissions worldwide. Not only would the equal rationing of a commodity (in this case carbon dioxide and other greenhouse gases) among all human beings be an extremely unlikely first for the species, but there are also countless different scenarios to reduce emissions. So, while the contraction and convergence idea helps to inform international negotiations, Australia's emissions target isn't tied at the hip to 450ppm, 550ppm or anything else. It has virtually no bearing on whether the world achieves such figures, and our national emissions target neither "locks us in to" or "saves us from" environmental destruction. Our emissions cuts, and the way we deliver them, are informed rather than dictated by science and economics. It's ultimately a political and moral decision about what constitutes a reasonable and affordable contribution to an effective global response.

Having recommended that the Australian government set a very low target if international negotiations faltered as he expected, Garnaut endorsed almost every loophole required for Australia to export its emission obligations and for our worst polluting sectors to offload theirs. He proposed generous compensation for emission-intensive, trade-exposed sectors. The biggest polluters should have most of their emission costs paid for by cleaner businesses and consumers until there was a more level international playing field – something that might take decades.

To help Australia achieve its target, Garnaut said that emissions trading should be integrated with developing nations such as Papua New Guinea and Indonesia, where vast quantities of carbon credits could potentially be generated through "reducing emissions from deforestation and degradation" (otherwise known as REDD). Around 17 per cent of the world's annual greenhouse emissions come from deforestation, mainly rainforests in Amazonia, South-East Asia and West Africa. Using the carbon-storage service that forests provide to fund their own protection is a worthwhile endeavour, but it's also fraught with danger. There's no guarantee the carbon "saved" through avoided deforestation will not "leak" elsewhere when timber, palm oil or cattle production moves to the next forest or country. Moreover, it enables emission-intensive industries (and national governments) to avoid cutting actual greenhouse pollution with the stroke of a pen. Unlike CCS, it doesn't require pipelines, underground reservoirs and technological breakthroughs. You can bury your greenhouse pollution in a Papua New Guinean rainforest by merely persuading the landowner to preserve his forest. You write a cheque, sign a contract and take credit for the emissions "saved."

It's also a cheap option: estimates by the World Bank and the Stern Review suggest the price of carbon saved in this way could be as low as US$1–3 a tonne, a tiny fraction of the carbon price likely to prevail without such credits. Even within Australia, "avoided deforestation" is a bargain-basement option for polluters. In 2006, for example, Rio Tinto Aluminium stitched up a deal in Queensland locking up 30,000 hectares to generate some 1.25 million carbon credits for less than $2 a tonne. Companies like Rio are eagerly eyeing similar deals offshore, but on a much grander scale. With some 8 billion tonnes of CO_2 annually that might be "saved," it's an irresistible bonanza for conservationists desperate to save forests, for polluters looking to avoid emission cuts, and for governments looking to make them on the cheap.

It was also irresistible for Ross Garnaut. Of Australia's emissions targets, he said, "We are talking about emission *entitlements*, not actual emissions." Paying others to cut emissions "would benefit both sides," he claimed. "The financial flows would benefit Indonesia and Papua New Guinea while Australia would benefit from access to low-cost abatement options." Matthew Warren at the *Australian* summed up the strategy:

> Rich countries such as Australia are going to need to rely on getting their hands on lower-cost greenhouse abatement opportunities in these developing countries if they have any hope of making the even deeper emissions cuts ... The only way ... will be to work with developing economies, starting with neighbours such as Indonesia and Papua New Guinea, to reap the low-cost credits that come from halting and reversing deforestation, which alone accounts for 20 per cent of global greenhouse emissions.

A scenario was emerging, and with it an uneasy sense of déjà vu. Howard had used a lessening of deforestation in his own country to hide emissions and remain on track to meet the Kyoto target. If Rudd accepted Garnaut's advice, Australia might achieve its new target by decreasing deforestation in other countries without making any "actual emission cuts" at home. If the targets were sufficiently unambitious and the potential credits offshore sufficiently abundant, Australia could even increase emissions at home while remaining "on track" with its Copenhagen target. The big question was whether Rudd would favour the same approach as Garnaut.

Most expected Kevin Rudd to be even less ambitious than Garnaut, on the basis that governments are more attuned than professors to lobbying and electoral danger. There were already signs that Rudd would underwhelm on climate change. Flush with cash, Labor's first budget directed relatively little to it: $2.3 billion over four years. This was much more generous than John Howard, but for every dollar spent on greenhouse programs the government was apparently content with close to fifteen dollars more being spent subsidising fossil-energy use. Rather than encouraging motorists to make more efficient transport choices now, the government favoured a billion-dollar plan that might lead to a small number of foreign-made but locally assembled "green vehicles" rolling out of Australian factories in a few years. To mix the messages to consumers further, Rudd flagged the possibility that motorists might be shielded from the impact of emissions trading on the price of petrol. Rather than moving away from fossil-energy dependence, the government dug deeper. "We have to find more oil," said Martin Ferguson, and to help keep oil prices lower it was time the G8 took a blow torch to OPEC, said Rudd.

Though there were occasional differences, the signs through 2008 were that Rudd and Garnaut were on the same page. The prime minister made clear in visits to Indonesia and Papua New Guinea that he was keen to "tie rainforest protection to emerging global carbon markets … to set a value – a real dollar value – on the carbon stored in rainforests." This, he told audiences in both countries, could become a big money-spinner.

Yet until the government detailed its preferred approach to emission targets, timetables, trajectories and the fine print of emissions trading, nothing was certain. Unlikely as it seemed, Rudd might set an ambitious agenda and block most of the loopholes being sought by the largest polluters. However, the prospects of this faded in July 2008 when the climate-change minister, Penny Wong, released the government's Green

Paper. For me, this all but sealed the carbon capture of Kevin Rudd. In keeping with the *Yes Minister* tradition of "always disposing of the difficult bit in the title" because "it does less harm there," the "Carbon Pollution Reduction Scheme" would not in fact reduce carbon pollution in Australia – it would actually rise, as I'll discuss below. The plan largely embraced the greenhouse mafia's plan to avoid emission cuts, and placed various obstacles in the way of the very transformation to a low-carbon economy that it promised.

Against Garnaut's recommendation, coal-fired power generators would be compensated for the decline in the value of their businesses. Investors who made dirty investments knowing carbon pricing was coming would be given a "once and for all" reward for doing so, and state-owned generators would effectively be fattened up for future sale. Against Garnaut's advice, transport would "initially" be carved out of the emissions trading scheme: although transport generates over 15 per cent of Australia's greenhouse pollution; although the average fuel efficiency of passenger vehicles hasn't improved since the early 1960s; and although the freight and logistics industry has been among the most fatalistic and lethargic sectors in the greenhouse debate.

Having rejected Garnaut's few hard lines, Wong proposed generous compensation for the biggest polluters. Some would get 60 per cent of their permits free, some would get 90 per cent – the dirtier the industry, the bigger the compo. On average, industries responsible for 37 per cent of Australia's greenhouse emissions would end up paying for one in every five tonnes of the carbon dioxide they generated. The biggest winner was the coal industry. Rudd was "a big believer in coal," and so was his government. They showed a seemingly boundless faith in carbon capture and storage. It could save coal-fired electricity generation in Australia, protect our coal exports, and it was a technology we could sell the world. It might even provide energy security through coal-to-liquids, which Rudd's energy minister said was going to "play a major role in Australia's energy future." A former coal-industry lobbyist was tasked with guiding the government's CCS strategy. There were large new subsidies for "clean-coal" R&D and a new Global Carbon Capture and Storage Institute. Lobbyists who had bragged to me in taped interviews that the Howard government had let them in to write Cabinet submissions and ministerial briefings were now back inside with the Rudd government in an official capacity. There would be no moratorium on building new coal-fired power stations, nor any time limit set by which new coal-fired generation needed to be "clean." With Australia's coal exports spiralling and none being used cleanly, NASA's James Hansen, perhaps the world's best-known and most widely respected climate scientist, had written to Rudd as the Green Paper was being drafted, urging Australia to exit coal mining and exports until CCS arrives on scale and to block any more conventional coal-fired generation. He was assiduously ignored. Labor's "light on the hill," it seemed, was a smouldering pile of coal.

With so many permits to be given away to a few big polluting industries, the task of cutting Australia's emissions would become commensurately harder for "non-assisted businesses" and consumers. When ABARE had advised the Howard government that carving out the biggest emitters would double the carbon price for everyone else, they had not factored in nearly as much compensation as the Rudd government now proposed. The pledge to offset the impact on petrol prices might sugar-coat the pill for consumers, but it merely shifted the cost onto other bills, such as electricity. Once the dust settled, the "non-assisted" might pay three times as much for carbon as they should.

Trebling the eventual carbon price paid by working families to protect Australia's worst polluting companies isn't a great look for Kevin Rudd. To hide it, various cosmetic measures were canvassed in the Green Paper, chief among these on the home front being the expansion of the forestry sector. The potential of this is significant but much less than recent scaremongering suggests. Some have claimed that 40 million hectares of prime agricultural land in and around Queensland's south-east will be turned over to new plantations, forcing farmers off their land, forcing food prices up and destroying communities. In reality, the potential of carbon-sink plantations is limited almost entirely to low-rainfall areas far inland, because these are the only places where the anticipated carbon prices enable such projects to compete against other land uses. Still, Treasury modelling suggests that around 5 million hectares of new planting might save on average 25 million tonnes of CO_2-e per year between now and 2050. This would offset around four large coal-fired power stations.

But the real action was overseas. Australia, it emerged, is banking on being able to buy *billions* of tonnes of carbon credits from developing nations over the next couple of decades. The Kyoto Protocol allows rich countries to buy credits by financing emission cuts in poorer nations under a program called the Clean Development Mechanism (CDM). As a result, a substantial amount of money has been transferred to fund highly worthwhile energy-efficiency and renewable-energy measures. Australia is not the only developed country looking to export emission cuts. However, these credits were only ever meant to be "supplementary" rather than a substitute for cuts by developed countries. Not only does it now appear that Australia wants to abandon the "supplementarity principle", but the credits it is mostly counting on come not from financing renewable energy but from hiding our pollution in the forests of South-East Asia. While Garnaut had suggested but not specified a limit on the use of such credits, the Green Paper merely said that they would not be included in the first couple of years of emissions trading, but that this would be reviewed later on. The clear long-term intention was to make the credits fully fungible well before 2020, so that they might count towards any target Australia accepts in Copenhagen.

It is unlikely that the penny will drop any time soon. The fine print around forest credits will probably be written well after Copenhagen. However, with deforestation in Papua New Guinea and Indonesia accounting for some 1750 million tonnes of CO_2 per year (around three times Australia's annual emissions), credits from these countries could enable Australia to meet cheaply almost any conceivable 2020 target by merely writing cheques. This "hollow logs" strategy should be cause for alarm. With these credits potentially selling for as little as one-tenth of the current price of carbon traded in the European scheme, there might be virtually no incentive to cut industrial emissions in Australia.

Of course, if Australia didn't intend to rely on carbon colonialism to meet its next international target, it would recommit itself to the "supplementarity principle." The Rudd government would stipulate that, while we support forest protection, we will make the vast majority of our emission cuts between now and 2050 here, not offshore. We'd support a separate market or global fund for forest credits so as not to compromise any constraint on industrial emissions. We might even make the biggest polluters pay for more than 20 per cent of their emissions and divert some of the money raised to reducing deforestation. Rather than counting such credits towards further emission targets, we'd treat them as an additional contribution.

So far, the Australian government has avoided saying any of these things, and Treasury's modelling in late 2008 revealed much about the likely approach. It reinforced the Green Paper's implication that Australia had no plans to cut *its* actual emissions by anything like 60 per cent by 2050 but would instead rely heavily on buying emission credits offshore. Treasury's modellers assumed that at least until 2020 these purchases would constitute up to half of our "mitigation effort." By 2050, it was envisaged, Australia would return its *actual* greenhouse pollution to 1990 levels rather than make any deep cuts.

As the end of 2008 approached, there was a sense of foreboding among many who had voted in 2007 expecting much better from the new government. After months of relentless lobbying, further backsliding seemed inevitable. The White Paper would be to the Green Paper what the Green Paper had been to the Garnaut Report – spectacular proof that the common denominator can be lowered even further. On the day of its release, it was positively Howardesque the way Kevin Rudd took charge and bumped his climate-change minister from her National Press Club platform in order to proclaim Australia was leading the world on climate change.

Rudd's rhetoric was utterly at odds with the policy being announced – as if the speaker hadn't seen the policy. Adopting the lower target flagged in the Green Paper and recommended by Garnaut, he said he would commit to a 5 per cent emissions cut relative to 2000 levels no matter what. Australia would accept a 15 per cent target only in "the context of a global agreement where major economies agree to substantially restrain carbon pollution and advanced economies take on reductions comparable to Australia." It was strongly in Australia's interests that the world pursue a more ambitious 450ppm objective, said Rudd, but Australia would not consider a stronger national target "consistent" with a 450ppm global objective until after 2020, by which time such an objective would almost certainly be unattainable. As Treasury's work had foreshadowed, the proposed targets didn't mean cuts in actual emissions – just our emissions *allocation*. The White Paper put no limit whatsoever on the use of foreign permits and credits: it would be open slather on exporting our emission cuts. Now, at least in theory, Australia could make almost all of its "deep cuts" by paying other countries to make them. Yet with a straight face Kevin Rudd said his package honoured his election commitment to produce "a blueprint for reducing carbon pollution *at home*."

Rudd then claimed that his 5–15 per cent target range for 2020 was more ambitious on a per capita basis than the EU's and president-elect

Barack Obama's commitments. Rapid population growth made Australia's efforts all the more impressive, so the story went: on a per capita basis, our 5–15 per cent target equated to a 34–41 per cent cut in "carbon emissions for every man, woman and child in Australia over that time, over the period 1990 to 2020." Yet the impression that greenhouse pollution was being cut by the actions of the Rudd government could hardly be further from the truth. Around two-thirds of the per capita improvement came from land-clearing cuts and forests planted before he came to office.

As neither the US nor the EU have such a fortuitous disguise for their pollution, a fairer comparison is between emissions per capita excluding land-use change and forestry. In 1990 Australia's emissions were 418 million tonnes, today they are around 575 million – it's the huge cuts in land clearing that conceal this 38 per cent increase in actual greenhouse pollution. By 2020, according to Treasury's projections, emissions under the Rudd government's 5 per cent target will be around 560 million tonnes (excluding land-use change and forestry). This is still 34 per cent higher than in 1990 and far less ambitious than EU and US proposals on a per capita basis.

The Rudd government's emphasis on per capita emissions is disingenuous for other reasons, too. It skates over the fact that Australia's per capita emissions are the highest in the developed world and would still be more than twice as large as the EU's by 2020. It ignores the reality that increased population is a policy choice this country has made: China's one-child policy could hardly be in starker contrast to the three-child policy championed by Peter Costello (one for Mum, one for Dad, and one for the country). Countries that use immigration programs, "baby bonuses" and other inducements to deliberately accelerate population growth *should* have to cut emissions per capita more deeply. Moreover, the climate cares only about total emissions accumulating in the atmosphere – this is what drives global warming, not per capita emissions. Kevin Rudd's tactic was the purest spin: focusing on per capita improvements to hide the reality that *actual* greenhouse pollution is not being cut overall.

Having set an emissions target that could be met without cuts at home, having removed all limits to exporting cuts, and having misrepresented this as a superior effort to that of other developed nations, all that remained for Kevin Rudd was to compensate Australia's leading emitters. The government implied that polluters would be forced to pay for *all* their greenhouse pollution. "Nobody gets a free ride," said Penny Wong. A glossy pamphlet called *The Carbon Pollution Reduction Scheme and You* went even further, saying "under the scheme, Australia's biggest polluters will pay for the pollution they generate." What it didn't say, and what Wong dared not say, was that almost all of these polluters would pay for only a tiny fraction of the pollution they generate – about one in every five tonnes on average. The rest of us would effectively pay for the other four.

The spectacular burden-shifting kicked off with $3.9 billion for coal-fired generators – once-and-for-all compensation for the fall in the value of their assets. When last I checked, Australia's military commitment in Iraq cost less than this. The White Paper accepted Treasury's view that there was very little risk of carbon leakage if industries were forced to pay for their emissions, then thoroughly ignored this advice. The threshold at which companies would be eligible to receive 90 per cent of their permits free remained the same, but the threshold for those receiving 60 per cent of their permits free was lowered. Woodside finally got its wish: LNG production would qualify for free permits. Although coal mining met the formula, the government was afraid of how it would look to give the coal industry free permits, so a $750 million abatement program was put in place instead.

The White Paper provided an *estimate of* rather than any *limit on* the number of free permits that would be given to eligible polluters. The government said it expected that by 2020, 35 per cent of permits would be provided free for eligible industries. It might even be higher if these sectors grew faster than the rest of the economy. In rough numbers, if those issued free permits used rather than sold them, a 60 per cent cut in overall emissions would only be possible if everyone else made cuts of more than 90 per cent. If the proportion of permits gifted to polluters rose to 45 per cent by 2020 (as the government expected it would once agriculture was included), a 60 per cent cut in actual emissions in Australia would be impossible even if the rest of the economy cut its emissions to zero. And compensation to big polluters would only be wound back in the distant future if other countries agreed to do likewise – an extremely bleak prospect, if global trade negotiations are any guide.

The whole scheme looked like it might cost rather than raise money: taxes might need to be raised to keep compensating everyone. The question was whether voters would realise what was happening amid the flurry of cheques. The White Paper set up a money-go-round of compensation payments so that everyone felt better off without appreciating who the real winners and losers were. Small businesses and community groups would share $1.4 billion to help them with the transition, and householders would be bombarded with small cheques. According to the White Paper, households faced an average increased "energy cost" of around $312 a year, but assistance would be coming to "fully meet" the extra impact on living costs for low-income households and to "help meet" the impact for middle-income households. All households would also "benefit" from the government's decision to offset the impact of emissions trading on the price of petrol and diesel. The only group of households not assisted would be the very wealthiest. Consistent with the government's approach, the householder assistance was designed as if to defeat the purpose of a carbon price. Increasing cash payments would merely erase the incentive for people to change their behaviour – and to the extent that some householders were to be "over-compensated," it might even increase emissions.

Deciphering what cheques were coming would be tricky: people had to compare their incomes and circumstances against all the changes. The White Paper included a long list of tables to help people in this near-impossible task, but nowhere were the relatively much larger payments being made to polluters on behalf of each household listed. Nowhere was there a table showing that a one-off $455 cheque would be written on behalf of every Australian household to prop up the value of coal-fired power stations for their mostly government and foreign owners. There was no mention of the one-off $93 cheque being written to the coal-mining industry in lieu of the free permits deemed politically beyond the pale by Cabinet. And nowhere was there a table showing that the equivalent of cheques worth on average $500 per household would be written each year between now and 2020 (and beyond) to buy free permits for Australia's worst polluting industries. There was no list showing that Rio Tinto would get over half a billion dollars annually in such cheques by 2015, nor that this figure would keep increasing. Moreover, nowhere was the net effect of this money-go-round acknowledged: to entrench the status quo, to dupe people into believing action was underway, and to export our emission cuts to the greatest extent possible.

It was a surrender to the same forces in whose interests John Howard had governed, but with one important difference. The question of *whether* emission cuts would occur was now gone, because, unlike Howard, Rudd *was* agreeing to take on obligations commensurate with a 60 per cent reduction by mid-century. The policy agenda had shifted markedly. How deeply and quickly Australia should cut emissions was still contentious, but *quantity* and *timing* were no longer the central issues. The big questions now were to do with the *quality* and *morality* of Australia's emission cuts: where the emissions were cut, who made the cuts, how the cost of the cuts was apportioned, and whether the answers to these questions would be made with the short- or long-term interest of the nation in mind. Rudd's agenda suggested that minimising the cost of cutting emissions was more important than the quality and morality of the cuts. If it saves a few bucks now, so be it: it doesn't matter that most of our emission cuts are made offshore, that our actual emissions don't fall at all, that our worst polluters are subsidised by everyone else to an ever greater extent each year, or that all the out-sourcing and burden-shifting denies the nation and future generations the opportunities and skills required to prosper in a carbon-constrained global economy. All that matters is that our emission cuts are made somewhere, and that they are made as cheaply as possible. Having failed to recognise this profound shift in the policy agenda, much of the environmental movement was left stunned by the White Paper – sucker-punched. Many had thought Australia merely needed to adopt a target between 25 and 40 per cent, not grasping the fading significance of the target itself and the rising importance of the loopholes and carve-outs.

The blood-and-thunder prophets were unsurprised by, and quietly satisfied with, the White Paper. They made their obligatory protests, mind you – this will still be a "stretch" said the Australian Industry Group; coal miners might not be able to compete against Mozambique and Mongolia without even more compensation, said the Australian Coal Association; the coal-fired power generators said the once-and-for-all hand-out wasn't nearly enough. Ross Garnaut attempted to distance himself from the government's position. He had recommended the 5 per cent target that Rudd ultimately adopted; he had said ambitious global action targeting 450ppm was unachievable and on that basis Rudd had ruled out ambitious targets until 2020; he'd enthusiastically backed outsourcing Australia's emission cuts in nearby forests; and he'd championed big handouts for the worst polluting sectors. Now, having drawn up the blueprints and built the foundations, he refused to occupy the finished house.

Most of the Canberra press gallery, meanwhile, were mesmerised by the political wizardry of the government. The long-term national interest, the economic and environmental implications and the palpable sense of betrayal felt by millions of Australians who expected better from Labor – these things were much less important than Labor's impressive political acrobatics. The whirlwind of cheques leaving everyone feeling no worse off, the deft management of expectations and the strategic positioning of Labor in the sought-after middle ground – these things were paramount. Fiscal churn on this scale was a thing of beauty. With a handful of exceptions (Tim Colebatch, Olga Galacho, Mike Steketee, Bernard Keane and a few others) most of the press gallery wore their election goggles too tightly: how cunningly Labor had fooled green voters with symbolic commitments in the 2007 election to hide the detail of a Howardesque policy; how cleverly Labor was using the issue now to split the Coalition and position itself for the 2010 election. Paul Kelly's take on Rudd's policy summed up the dominant attitude:

> Rudd will define Labor forever as the party that acted on climate change ... Rudd is a green Howard. He has turned climate change into a magic pudding. It is a work of political genius that would make Howard proud ...

To the Machiavellian minds of some commentators, emulating Howard on climate change was a good thing, and misleading the electorate so spectacularly was not just clever, it was to be encouraged. As Janet Albrechtsen wrote:

> Every now and then you have to be grateful when you discover our political leaders have told a deliberate, calculated lie ... his pre-election claim that climate change was the great moral issue of our time, and demanding that Australia lead the way, was what Winston Churchill would call a terminological inexactitude: a whopper, a piece of bare-faced duplicity of epic proportions. But thank goodness Rudd and his colleagues deceived us.

One can only imagine the opinions of those without a newspaper column to express their views: the millions of Australians for whom the last thing they wanted on climate change was that Kevin Rudd should mimic John Howard. To most of these people, the *quality* and *morality* of Australia's emission cuts does matter a great deal.

Despite their public differences, for both Kevin Rudd and Australia's greenhouse mafia this is a win-win outcome. Rudd appears slightly greener than the Coalition, but not as zealous as the Greens, and public friction with big emitters belies his lack of ambition and his embrace of polluter carve-outs and loopholes. With so many industries climbing onto the life raft, you'd expect it might sink – the aluminium industry even warned as much – but Labor seems intent on accommodating them all. The man who said so often in Opposition that Australia should be more than China's quarry, who said the quarry export strategy wasn't working, and who raised so many hopes, has dug us even deeper.

I'm left wondering what it will take for Australia to change course as time ticks away and emissions move onwards and upwards. Was Labor's response really pre-ordained by the political and economic circumstances? Was this "huge fiscal churn" *really* "the only way Australia was ever going to price carbon?" Out-greening the Coalition might help secure a second term, but Labor hasn't found the "sensible middle ground" or "got the balance right." They have merely chosen the ground next to the Coalition. Now Australia has two fossilised parties – more "proud economic conservatives," more of coal's "happy clappers," looking to hide the problem under a rock, in a tree or in a policy-paper title. I think Kevin Rudd has underestimated how determined Australians are for genuine and urgent action on this, the biggest issue of our time. In doing so, he contributes to a growing vacuum in Australian politics.

THE PROBLEM WITH "CLEAN COAL"

No matter what happens in 2009, Australians will still be conscripts on the wrong side of a "coal war" with climate change, a costly and disastrous proxy war on behalf of our coal industry. The industry may prevail, but we will lose, as will the planet – it is merely the extent of the loss that is uncertain. Some call for a "War Cabinet" to fight climate change, perhaps not noticing the several we already have. Successive state and federal Cabinets have fought on coal's behalf in the war on climate change, believing coal burning has a bright future. They dare not consider the possibility that they have backed the wrong side. Rather than glance at that Medusa-like prospect for fear they might turn to political stone, they have turned to anthracite and lignite, oblivious of the consequences. Tim Winton might just as well have been describing the nation's fight to protect coal when he recently wrote of his home state of Western Australia that the "economy and mindset are bound up in an endless war against nature." Ultimately, however, the Australian people can only prevail if their country changes sides. The sooner we prepare for that, the better for the global environment, and the better for Australia.

There is more to our strong alliance with the coal industry than historical and psychological attachment to the quarry; there are more pragmatic reasons. Under the accounting provisions of the UN's climate-change convention, the world has agreed that greenhouse emissions are attributed to the nations where they occur. To the extent that any "extended producer responsibility" is accepted for our coal exports, it ends once the ships leave our ports. Thus, Australia is responsible for less than one-quarter of the CO_2 emissions generated by its coal because around 80 per cent of it is exported. Coal is still a major source of domestic emissions, mainly because burning it generates most of our electricity, but by far the greater contribution of Australian coal to climate change accrues in other nations and is ignored relatively easily by us. As the world's largest "coal mule," this ought to present a moral challenge to a country that readily points the finger of climate-change culpability at the USA, China and India (the world's largest coal users). Yet it is one that Australian political leaders continue to dodge.

Aside from accounting convenience, it's politically attractive to keep coal exports off the table; we're in so deep. Governments have a financial stake in coal mining, burning and exporting: they own generators and rely on exploration-rights revenue and mining royalties; their railways, ports and other infrastructure depend on coal and its expansion. On both sides of politics, senior ministers, their predecessors and their mentors have made many of the decisions that dug Australia deeper into coal. As well as being personally invested, they have close friends and colleagues lobbying on behalf of coal companies or sitting on their boards. In the bureaucracy, there's a similar dynamic. A career spent making decisions that assume coal expansion in spite of climate change may not constitute a financial interest, but it confers a degree of ownership of the status quo and it fosters resistance to change.

For business leaders whose working lives were spent persuading cautious multinationals to bet on coal's expansion in Australia, and union leaders with thousands of members feeling under siege because of climate change, fighting the "coal war" seems the only conceivable option. It is unthinkable for such people to consider changing course, let alone changing sides – the humiliation and animosity are unimaginable. There are other unwelcome considerations. If Australia turned back now, there would surely be lawsuits from multinationals arguing that the government has no right to interfere after having sent such receptive signals to investors for so long. If Australian withdrawal from the coal trade pushed up coal prices, there would surely be diplomatic fallout, too – an outcry as coal export customers say we have no right to deprive them of essential energy resources. Various national governments that have invested in coal mining in Australia might retaliate. Critics would argue against singling coal out for special attention. After all, if we're not going to export coal until it's used cleanly, why should other countries send us oil, cars or plasma TVs until they're also used "cleanly" here? Why should there be a different rule for coal and LNG? The latter is cleaner, but it is still a fossil fuel.

The arguments are predictable, as if hundreds of millions of tonnes of coal were as innocuous an export as hundreds of millions of Kylie Minogue CDs. In the end, however, coal is a special case. There is still enough of it in the ground to meet projected global demand for hundreds of years, whereas the rising long-term price of oil is already starting to give rise to alternatives. As for the emissions generated by the use of most other energy-intensive products (plasma TVs, game consoles, fridges and so on), these can be cut to zero by switching to green power – an affordable option available to almost any user today. As for LNG, the emissions associated with our exports to 2030 are projected to be perhaps one-sixth of those caused by our coal exports – it is not in the same league. So coal *is* different. It is cheap, plentiful and inherently dirty. Coal is also different in that using it "cleanly" relies on a silver bullet: massive-scale carbon capture and storage infrastructure that must be virtually completed, not merely underway, ten to fifteen years from now, if there is to be any hope of avoiding catastrophic climate change. Given that the prospects of such infrastructure being in place are nil, as I'll discuss below, it's unthinkable *not* to single out coal. The urgent timetable for emission cuts is such that we simply can't keep burning it. The best summary of the situation, by James Hansen, is worth quoting at length:

> The only realistic way to sharply curtail CO_2 emissions is to phase out coal use except where CO_2 is captured and sequestered … if the world continues on a business-as-usual path for even another decade without initiating phase-out of unconstrained coal use, prospects for avoiding a dangerously large, extended overshoot of the 350ppm [CO_2] level will be dim … Present policies, with continued construction of coal-fired power plants without CO_2 capture, suggest that decision-makers do not appreciate the gravity of

> the situation. We must begin to move now toward the era beyond fossil fuels. The most difficult task, phase-out over the next 20–25 years of coal use that does not capture CO_2, is Herculean, yet feasible when compared with the efforts that went into World War II.

In a recent letter to Barack Obama, he went on to say:

> This is the sine qua non for solving the climate problem … Coal is responsible for as much atmospheric carbon dioxide as the other fossil fuels combined, and its reserves make coal even more important for the long run. Oil, the second greatest contributor to atmospheric carbon dioxide, is already substantially depleted, and it is impractical to capture carbon dioxide emitted by vehicles. But if coal emissions are phased out promptly, a range of actions including improved agricultural and forestry practices could bring the level of atmospheric carbon dioxide back down, out of the dangerous range … continued construction of coal-fired power plants will raise atmospheric carbon dioxide to a level at least approaching 500ppm (parts per million) … a conservative estimate for the number of species that would be exterminated (committed to extinction) is one million. The proportionate contribution of a single power plant operating 50 years … would be about 400 species! Coal plants are factories of death. It is no wonder that young people (and some not so young) are beginning to block new construction.

Coal could hardly loom larger in Australia's emissions equation. In total, the coal we burn at home and export generates close to a billion tones of carbon dioxide annually – more than 1 in every 30 tonnes generated globally from fossil-fuel burning. For a country with 1 in every 313 of the world's people, this is polluting well above our weight. Ships laden with a single load of Australian coal can each carry the annual emissions equivalent of 25,000 cars. Our exports alone generate almost 700 million tonnes of CO_2 annually, which is almost 15 per cent more than Australia's entire annual greenhouse-gas emissions. Yet currently our coal exports are projected to nearly double from 231 million tonnes per annum in 2004–05 to 438 million in 2029–30. As a share of the global total, our *domestic* emissions are falling slowly, but our overall contribution to climate change is increasing. Even if we cut emissions in Australia to zero tomorrow, the projected increase in coal exports erases the environmental saving – by the mid-2030s our carbon footprint would still be roughly what it is now.

Thus, no credible Australian response can ignore coal exports and there are only two ways to change the equation: either CCS saves the coal industry, or we phase out our role in coal mining and coal burning. The former is simply impossible, the latter merely unthinkable. This is obscured at present because the coal industry has defined clean coal, and the definition of success, on its own terms. Between the coal companies and governments, hardly a month passes without the industry or its proxies announcing another "clean-coal milestone." For each new phase in the few pilot plants demonstrating CCS, for each new round of funding, for each cheque written by a different level of government, ribbons need cutting and the cameras need another grip 'n' grin moment. It is all proof that Australia has "skin in the game," as the prime minister says we must. Based on my rough calculations, a new pro-CCS media release is issued for about every 100 tonnes of CO_2 actually captured and stored. The spectacular growth industry is not clean coal, but clean-coal PR.

To reinforce the impression that progress is being made with CCS, the coal industry runs a massive PR campaign to keep the image of "clean coal" as clean as it can. In the USA, coal-industry-funded front groups spent tens of millions in the run-up to the recent presidential election. The campaign used the same coloured backdrop in almost all its advertising – "pie-in-the-sky blue," one might call it. I'm told that the industry's market research found that consumers subliminally associate the colour with clean technology. The advertising actually features blocks of coal with electric cords plugged straight in – as if the burning-and-emitting part can be magically bypassed! One ad features the slogan "I believe" seventeen times in sixty seconds, as if enough faith and repetition will miraculously make "clean coal" happen.

The Australian coal industry has recently launched a similar re-branding campaign. The appeal to nationalism is similar, faith in American ingenuity is replaced by Aussie know-how, and other changes have been made to make the message more locally relevant. Rather than appealing to our religious fervour by telling us to "believe," the Australian coal industry deftly shares the load by urging people to do *their* bit: to "Flick the switch" and "Turn it down." The industry is spending over two million dollars on a new website (the biggest of its kind in the world) and an ad campaign by the same people who came up with the Kevin07 campaign. The local advertising blitz and the new Australian Coal Association logo both feature the ubiquitous poll-driven sky blue. Inconveniently, however, Australian consumers are suspicious about the "clean-coal" brand. So, to combat punter wariness, a new market-research-driven term is now in play: "NewGenCoal" feigns concern about the planet's future while treating a new generation of coal-fired power stations as a fait accompli. The Rudd government has done its bit for the re-branding campaign, too. The government now says that "there is no alternative" to a new generation of coal-fired power stations, it has changed the name of what was the National Clean Coal Fund to the National Low Emissions Coal Initiative, and half a dozen or more government employees from CSIRO and Geoscience Australia feature on the coal industry's NewGenCoal website.

Most demonstration projects focus on one part of CCS: either post-combustion carbon capture, or CO_2 transport, or geosequestration. They do not combine all the elements, let alone prove commercial viability. Almost all are heavily subsidised by government, and even then some of the biggest projects have been shelved. Just a year after it was announced with great fanfare, the Rio Tinto/BP coal-to-hydrogen project, touted as "Australia's biggest contribution to developing clean-coal technology," was abandoned. The Shell/Anglo Monash Energy coal-to-liquids project, which has hypnotised politicians since 2006, was also shelved recently. Collectively, if all goes well, the remaining carbon capture and storage projects expected to be up and running in Australia over the next few years might save 100,000 tonnes of CO_2 annually. It sounds impressive, until you realise that this is the same as a week's emissions from just one large coal-fired power station, or the equivalent of taking all the cars off the road in, say … Tamworth.

As evidence of the global momentum towards cleaning up coal, one carbon-lobby-funded research centre lists sixty-one operational or planned demonstration projects around the world. This too sounds impressive until you realise that forty-eight of these projects are in the USA, Australia and Canada. None whatsoever is listed from India, Indonesia, Russia or South America. Of the two projects listed from China, only one relates to coal and it concluded in 2004. However, there is another Chinese CCS project of note that is not on the centre's list and it is enthusiastically funded by the Australian government, in part so that prime ministers can visit it with a great fanfare when in Beijing. If all goes well, that particular "milestone" project will capture 3000 tonnes of CO_2 a year – the equivalent of taking all the cars off the road in Nimbin. There simply is no grand push for CCS in the countries where the fastest growth in coal-fired power is occurring.

International estimates suggest no commercial-scale deployment of CCS plants before 2020 in developed countries, and in developing countries not until at least 2025. The International Energy Agency reckons that it might take a carbon price of A$100 a tonne before CCS becomes viable (much higher than for renewable options such as wind and solar-thermal power). Perhaps not surprisingly, the IPCC doesn't expect carbon capture to contribute much until well into the second half of this century – too late for the climate. Modelling by the Australian Treasury suggests that CCS won't be deployed here until 2026, and not until 2033 under the 5 per cent target scenario now adopted by the government. Even our coal-fired electricity generators say they doubt that CCS will be viable on anything like the scale required in time. The head of the National Generators Forum, John Boshier, told the ABC's *7.30 Report*:

> I think we all felt a few years ago that clean coal was doable and was a great option for Australia. We've got a lot of coal in Australia. We're now worried about how long it will take and how much it's going to cost on the scale that we're talking about. We shouldn't kid ourselves it's going to be available in the near term. We think that 2020 is the earliest it can really be commercialised.

In the meantime, based on current projections, well over 1000 conventional coal-fired power stations will be added in the developing world. That would add around 8 billion more tonnes of CO_2 annually from coal burning alone (more than a quarter of the current global total produced by fossil-fuel consumption). Under such circumstances, there is simply no way that global emissions can peak by 2015, or even 2030, let alone fall rapidly.

When "commercial-scale" power stations are finally built with CCS, they're likely to cost twice as much, use a third more coal in order to power the carbon-capture process, and produce electricity that is twice as expensive. In the meantime, proponents argue that new conventional coal-fired power stations should be permitted so long as they're "CCS ready." Such facilities would be "CCS ready" much as I am "winning-lottery-ticket ready" – that is, it's a nice thought but there is little chance of it actually happening. Installing new CCS technology in old stations is likely to be very expensive, which is why some generators have publicly admitted that they expect it to make more sense to build a new power station than to retrofit an existing one. Pointing to the "enormous energy and cost penalty" involved, a power-company executive from Stanwell Corporation told a recent Australian parliamentary inquiry that "unless there is an enormous breakthrough in science, the post-combustion capture technologies would probably send you down the route of thinking, 'I might build something brand-new instead.'"

In early 2008, the US government abandoned its investment in FutureGen, the world's largest "clean-coal" project. Today, the Integrated Gasification Combined Cycle technology that would have been used by FutureGen is dismissed as the "Rolls-Royce" of carbon capture and storage. Alternatives, like CS Energy's Oxyfuel process being demonstrated at Callide in Queensland, are promoted as offering results much sooner. Yet even if we go with the Holden Barina rather than wait for the Rolls-Royce, there is no prospect of enough being done in time. By 2011–13, CS Energy plans to sequester up to 27,000 tonnes of CO_2 annually at Callide. The emissions saved are about 1/150th of the annual emissions coming from the company's new Kogan Creek coal-fired power station. If all goes well by 2011, Australia would then have to upscale what is proposed at Callide 7000 times just to deal with its coal-fired electricity emissions. To deal with current coal emissions globally, the world would have to buy the Holden Barina of carbon capture more than 400,000 times over, and perhaps a million times over to deal with the coal-fired emissions projected for 2030.

With coal-fired power use projected to double over the next few decades, there is simply no prospect of CCS keeping up. That is why senior officials at the International Energy Agency say half of China's coal-fired power plants, at most, will be fitted with CCS in our lifetimes. It's as clear an admission as you're likely to see that CCS cannot clean up coal in time. Even this best-case scenario is pie in the sky. Not only would almost every new coal-fired power station need to be built with CCS, but thousands of existing power stations would need to be retrofitted at enormous expense. A web of pipelines linking power stations and storage sites would have to criss-cross some of the most densely populated parts of the planet. If trucks were used instead of pipelines to transport CO_2 to storage sites (as is happening in various projects), close to 2 million trucks would be added to the world's roads by 2030 (enough to stretch right round the globe end-to-end). The infrastructure required for CCS would be comparable to that of the global petroleum industry, only it would need to be built in a fraction of the time – and all to dispose of a waste that can be avoided in the first place with cheaper technology, much of which is already available. It would cost a fortune and arrive too late: the most expensive exercise in futility the world would ever have seen.

So when we speak today of whether CCS will work or not, how we define "work" is all important. In Australia and elsewhere, this has been reduced to the question of whether a particular technology might one day stack up commercially somewhere. The prime minister murmurs that the International Energy Agency reckons CCS will be economically feasible with a $20–40 carbon price and rests his case. In reality, nothing is commercially feasible everywhere – it depends on the circumstances. I'll be very surprised if CCS is not viable in some Australian locations in the next twenty years. However, we should not confuse this with CCS "working." When we ask whether CCS will "work," what really matters is timing and scale. To work well enough to save the 20 per cent of Australian coal used domestically, CCS has to be retrofitted to most of our existing coal-fired power stations and used in any new ones.

Even if all emissions from coal were dealt with in Australia, CCS would not have begun to work well enough to save the four-fifths of production destined for export markets. It would need to be retrofitted at hundreds of locations around the world where Australian coal is burned or would be in future – in power stations and steel mills over which Australian governments (and coal exporters) have no jurisdiction whatsoever. If emissions are to peak globally by 2015 to avoid the worst impacts of climate change, this needs to happen very soon. To complete a global phase-down of coal emissions over the next twenty to twenty-five years, which scientists like James Hansen say is essential, the process has to start even sooner in countries such as Australia.

Yet for all Kevin Rudd's talk of having "skin in the game" and the coal industry's hype about making a "fair and reasonable" contribution to CCS, very little money is being spent in relative terms. The $0.20 per tonne levied under the industry's Coal 21 Program is on average less than a thousandth of what each tonne of coal is expected to fetch in 2008–09.

The Australian government, meanwhile, has much more skin in expanding coal exports than in cleaning up their use. The National Low Emissions Coal Initiative is worth just $90 million a year between now and 2012. More is being spent by the Rudd government in the Hunter Valley alone on expanding railway infrastructure to double coal export capacity. Up in Queensland, the federal environment minister, Peter Garrett, has approved the Wiggins Island Coal Terminal, which will double the coal exported from Gladstone. All up, the extra coal exports from these two regions will generate almost as much greenhouse pollution annually as Australia's national total. And for all the hype and taxpayer-funded help, the Australian coal industry's aim is to have the equivalent of just one large clean-coal power station installed by 2020. The same shortfall in investment is evident around the world (and made worse by the continuing growth in conventional coal use). The G8 has set a target of a mere twenty "industrial-scale" CCS projects by 2020. It is far from certain that this will be achieved, but, assuming a generous definition of "industrial scale" and assuming all goes well, it is less than one-hundredth of what's required to deal with emissions from existing coal-fired power stations. That leaves an inescapable conclusion.

If four-fifths of our coal is exported and there's no prospect of it being used cleanly in time, what we have here is an industry whose products are still in demand but whose "social licence" must be withdrawn: a sunset industry by moral and environmental necessity. The choice facing Australia is what type of sunset that will be: a long one that fuels catastrophic climate change, or a short one that fights it. It is undoubtedly one of the biggest decisions this country will face.

Australians can indeed keep pretending that coal can be cleaned up in time, ignoring the damage caused and delaying the inevitable adjustment. This approach assumes that the economic benefits of coal justify the environmental risks of being wrong about CCS. Yet the equation is much less persuasive than we are led to believe. In 2008–09, thanks to a trebling in the price of coking coal and a doubling in the price of thermal coal, the coal industry might generate close to 4 per cent of GDP. However, coal prices have now plummeted to such an extent that the industry may well fall in value next year by a third – business commentator Terry McCrann says by more than half (but expect the industry to keep using the 2008–09 figures in its PR for years). While coal-mining jobs have increased in recent years to 30,000, fewer and fewer workers are required for every tonne of coal mined, so job growth in the sector is likely to be modest at best.

Today the majority of our coal goes to Japan, South Korea and Taiwan, but most of the projected growth in demand is from developing countries, especially China and India. Some say Australia has no choice but to dig deeper because the world depends on countries like Australia retrofitting China's coal-fired power stations with CCS. They argue that China has no choice over the next decade but to add a million megawatts of coal-fired capacity – the coal-fired power equivalent of twenty-five to thirty Australias. It beggars belief that people see this as a desirable outcome, given the enormous and irreversible mess it leaves China and the world in. But some people can't see the clouds for the silver lining. All they see is a win-win situation in which we save our coal exports and CCS technology becomes our next boom industry.

The CCS boosters insist that China cannot afford to use anything but coal, as if it is pre-ordained in five-year plans not yet written by the Chinese government. The capital costs of renewable alternatives are beyond them, we're told, not to mention the more expensive electricity generated. Yet China will probably eclipse the United States as the world's biggest economy by 2040 and it recently found around A$880 billion for a stimulus package – a sum more than sixteen times the combined value of the two stimulus packages announced by the Rudd government. China has also subsidised petroleum and coal-fired electricity prices for years. If just half of this money was spent on clean energy, it would go a long way towards removing the need for the many hundreds of new Chinese coal-fired power stations we are told are inevitable. If ever there was a country that could achieve just about anything it set its mind to, it is China. The one-child policy has already prevented the greenhouse-gas emissions of an extra 400 million people. In truth, China *can* afford to grow cleanly, and it must if we are to have any chance of responding effectively to climate change. We can and should help, but to export false hope about CCS does neither China nor Australia any favours.

For Australia, the economic benefits of peddling coal and false hopes are simply not worth the environmental costs. In the unlikely event that coal prices returned to boom-like levels and stayed there, Australia might earn $2 trillion between now and 2030 by exporting around 8 billion tonnes of coal (with most of the profits sent offshore). The vast majority of the 21.6 billion tonnes of CO_2 generated wouldn't be captured or stored: it would end up in the atmosphere. Conversely, if Australia's coal exports were phased out, the economy would not suffer nearly as much as we assume. A coal phase-out would take time, so not all of the potential coal-export revenue would be lost, and some of the losses would inevitably be offset by new economic activity (such as expanding renewable-energy capacity). Recall that coal only generates around three dollars in every hundred, and that more than twice as many people work in McDonald's as in coal mining. Our economy would recover both the dollars and the jobs. It would take one to two years to recover every cent of coal-export revenue foregone between now and 2030; GDP would double by 2031 instead of a year earlier. Employment would recover, too, just as it has bounced back after the loss of almost five times as many manufacturing jobs since the mid-1990s as coal mining provides today. It's a serious economic decision for this generation, but one that would dramatically reduce the carbon liabilities we bequeath to subsequent generations.

Australia is in a unique position to help the world avoid disaster by withdrawing from the coal trade. The main reason for spiralling emissions from coal is not rising energy demand, which can be met in various ways, but the availability of cheap coal internationally. Though traded coal accounts for only about a sixth of world consumption, it underpins expanded coal use, especially in power generation and steel production. Much of the global trade is run out of Australia, with four of the largest coal multinationals based here (BHP Billiton, Rio Tinto Coal, Xstrata Coal and Anglo Coal), and others like Peabody, Shenhua and Mitsubishi operating here too. The international coal trade is controlled by a few nations: ten of them control 98 per cent of the global trade, and Australia is by far the largest with a 30 per cent share. And by 2030, the US government's Energy Information Administration expects the coal trade to grow by more than 40 per cent, with Australia accounting for more than half of that growth. Imports by India are projected to increase ten-fold; and while China will still produce most of its own coal, its imports will more than double. In short, thousands of new power stations and steel mills depend on a handful of nations – most notably Australia – supplying coal in ever greater quantities. Our coal industry might look to downplay their significance, but the CO_2 emissions generated by our coal exports are on track to rival those generated by Saudi oil exports today. We have a bright future as one of the world's leading "carbon mules," if we want it.

Australia can avoid this fate by committing to a short sunset on coal supply and use where carbon emissions are not captured. I'm not suggesting that we "close down the coal industry by next Thursday," but that, rather than expanding coal mining, we should plan to phase it down over the next five to ten years so that no Australian coal is being used without CCS by 2020. Where Australian coal sold abroad is not used cleanly within five years, and no credible plan exists to ensure it is by 2020, we should phase out supply. There should also be a moratorium placed on any new coal mines or conventional coal-fired power stations. Any costs arising from legal action against the phase-out plan should be funded either through a levy on the coal industry or by emissions-trading revenue currently earmarked to compensate emission-intensive industry. This should also fund adjustment costs facing affected communities and workers.

Some will say "leave coal's future to the market," and perhaps they'd have a case if plans for emissions-trading markets were sufficiently ambitious, or if coal was just any old commodity. Among other things, my proposals will be called futile by the coal industry, which will argue that others will fill any gap Australia leaves in the coal trade. Resorting to the dope dealer's defence, however, only highlights the culpability of the few companies dominating the trade. For it is the same companies that dominate Australian coal production that would seek to fill the gap from their mines elsewhere.

The aim of such a withdrawal should not be to impose a harmful economic shock. If such a phase-down happened gradually, it would have a significant (but manageable) impact only on our largest coal customers: Japan, Korea, Taiwan and the EU. There wouldn't be a large short-term impact on those countries with large domestic supplies: China, India, the USA and Indonesia. However, these countries would not evade the odium around coal that would follow from a phase-out of Australian exports.

The gradual withdrawal of a third of the world's traded coal would make coal more expensive. In theory, the higher price would make more mines commercially viable. In practice, however, the loss of the world's largest exporter would create substantial uncertainty about the long-term security of supply in the coal trade. Producers and consumers alike would wonder who might follow Australia's lead. Countries without large domestic coal supplies would become much more cautious about taking the coal trade for granted. The same would apply to potential coal and coal-fired electricity investors.

In the longer run, as coal becomes less reliable and more expensive, cleaner choices will become more economically attractive. Ultimately, however, the decision to withdraw from the coal trade is a moral one. It recognises that not everything that is profitable is desirable, and sees dealing with our greatest carbon liability as being in the long-term interests of the nation and the world. If Australia did announce a coal phase-down plan, it would be one of the most important symbolic moments in the global struggle against catastrophic climate change. It might well begin what Robert Manne has called a "benign domino effect", by setting a very powerful example of national altruism for the benefit of all.

Australia should not stop at planning its own withdrawal from the coal trade – it should encourage others to do the same. The Australian federal police commissioner, Mick Keelty, recently said that climate change would be this century's biggest security threat, even if only some of the expected impacts occur – bigger, he implied, than the "War on Terror." He's right, and there's no good reason why we should apply lesser standards. We don't let Australia become a terrorist haven because other countries might – we lead by example. Similarly, we shouldn't allow Australia to remain a pollution haven because other countries might. In the global war on climate change, Australia's biggest contribution to an effective response would be to exit the coal trade. Nothing else would give us the same credibility or leverage internationally, and it would place pressure on both coal users and other coal exporters, including Indonesia, South Africa, Venezuela and the USA. In choosing to fill the vacuum left by Australia, they would knowingly (not to mention conspicuously) increase their contribution to climate change, along with their economic exposure, as it becomes increasingly clear that CCS will not save coal in time. Conversely, a joint agreement between these countries to withdraw from the coal trade on similar terms would be the biggest contribution to reducing global emissions thus far. A handful of nations could, together, help to pull the world back from its rush into the climate abyss, and Australia could lead that effort.

At present, all this is not just unlikely, it's unthinkable. Our political, business and union leaders, and even some prominent environmentalists, apply a different standard to Australian coal. The "precautionary principle" – that where there is significant uncertainty, we ought to err on the side of caution rather than risk severe harm – is abandoned. Instead we believe that "clean coal" just has to work. Believing that the coal trade is "too big to fail" hides both the possibility of failure and our responsibility for it. We readily assume failure when it comes to global negotiations to cut emissions, yet in spite of all the evidence staring us in the face, we won't countenance the idea that CCS will not deliver in time. In quintessential Lucky Country style, we have no Plan B at all (except perhaps nuclear, which is also too little too late). The United States government had contingency plans for war with Great Britain as late as 1939, yet we have no plan for a sunset on coal, which is infinitely more conceivable.

Instead we're digging up, shipping out and burning as much as we can. At home, we're digging coal communities deeper into trouble and salving their worries with snake oil about CCS, making the eventual adjustment that much harder. We're handing over stewardship for Murray-Darling tributaries, groundwater resources and drinking water catchments to coal miners whose practices have literally cracked the river beds and polluted a host of rivers already. We're investing billions on infrastructure to remove the obstacles and "bottlenecks" hindering coal exports.

Under such circumstances, talk of quitting coal is anathema – "un-Australian." It only comes from people who "hate the coal industry" or are motivated by "green religion," or both. That's the line of both major political parties, the mining industry, affiliated unions and many coal communities, and their angst is understandable. Yet the satisfaction derived from dismissing calls like mine is a painkiller, not a cure, and it soon wears off. I don't hate the coal industry, I'm not looking for another religion, and until quite recently I too thought coal's demise was unthinkable. I value the mining industry's contribution to Australia's prosperity – it underpins my concern about the quarry's carbon liabilities. As for coal, I don't deny that CCS might be commercially viable in some locations under some circumstances later this century. I very much doubt that coal mining and use in power generation, steel-making and other applications will ever die out. However, the worst impacts of climate change can only be avoided if the coal industry becomes a cottage industry as compared to today. If CCS and other advances one day enable it to contribute on a large scale without doing environmental damage, well and good. The coal is still on hand rather than doing great harm in the atmosphere.

We should not be swayed by the predictable protests of coal-mining companies and their biggest customers. It is up to governments to make hard policy decisions like this one, not BHP Billiton or Rio Tinto, not Anglo or Xstrata, or any of the other blood-and-thunder prophets. Their job is to make hay while the sun shines and to extract the best possible deals from government along the way. In spite of their howls of outrage at any plans for phasing down Australia's role in the coal trade, rest assured that they would adapt. Around 75 per cent of Australian coal is produced by diversified multinationals that would want to continue operating here. Of course they'll talk "capital flight," but they wouldn't flee Australia and allow others to profit from our iron ore, gold, copper, nickel and other commodities. They have quietly made contingency plans to ensure that they prosper in the event of policy decisions like this one. The largest coal consumers have also made such plans, and they too would adjust.

As it always has, the rush will move on. The supercycle would still be there to enjoy, for most of those same companies and for Australia:

consider the 350 million Chinese still likely to move to the city over the next two decades, the 50,000 skyscrapers that will need building. Our quarry would remain extremely valuable after a sunset on coal, only much cleaner. The resources sector would remain an important contributor to exports, employment, tax revenue and regional communities. The ASX would still be overweight with mining stocks, and our retirement incomes would be safe.

And what of the coal miners themselves? A phase-down of Australia's role in the coal trade buys them time to adjust, and it provides government with ample opportunity to assist in that process. Fortunately, in many areas now dependent on coal, there's significant potential in geothermal and solar-thermal power generation and in coal-seam gas production. Many coal workers may find much cleaner mining and energy jobs closer to home than they can currently imagine, and their communities may find a shift to these industries a much brighter prospect than the siege defence of "clean coal" into which they have been conscripted in recent years.

Sooner rather than later, Australians must think the unthinkable. We cannot expect the world to clean up its act unless we deal with our biggest contribution to the problem. With International Energy Agency graphs showing coal use trending torpidly onwards and upwards in future as the world's dominant energy source, it's easy to think continued reliance on coal is preordained. It's easy to overlook the catastrophic environmental consequences embodied in the graphed scenarios, and it's easy not to notice how consistently organisations like the IEA have underestimated the potential of renewable alternatives. Yet in reality, nothing is preordained. Try as people might to graph the future, coal's future will be decided in Canberra and a few other national capitals, and we must hope the political will can be found to do the right thing.

In truth, we will not be dealing with climate change as a nation until we deal with the carbon liabilities we export to the world. Tim Flannery stated in late 2008 that being willing to give up $50 billion in coal-export revenue was the sign of a "true lunatic." "Gouging into the earth is our business," he said. "That is the fact of our economic situation." To my mind not only does this reinforce how deeply our quarry vision is entrenched, but it insults lunatics everywhere. Banning child care, or television, or pubs, or the automobile would cause far greater economic damage, not to mention civil strife, than withdrawal from the coal trade. A far greater lunacy than contemplating coal's demise is continuing to export it in ever greater quantities knowing it will not be used cleanly in time on any meaningful scale.

Many Australians today share the view that coal exports are sacrosanct, but how long until that changes, I wonder? When do we decide that in spite of a handful of pilot projects and lots of PR, almost all of our coal won't be used "cleanly"? Is it still true lunacy to demand that coal be used cleanly or phased out in 2015, in 2020, or in 2050? How much more do we add to our export pile before it becomes untenable? Another 100 million tonnes, another 250 million? How many coal-laden *Pasha Bulkers* need to wash up on our beaches in cyclonic winds before we call time? How much deeper will we dig before then? Until coal royalties are 25 per cent of Queensland's tax base; until Beijing owns half of Rio Tinto along with many of our coal mines? Must we again wait for other countries to act before we contemplate withdrawing from the coal trade, and if so which countries: Indonesia, South Africa, Mongolia or Venezuela? Sooner or later this "true lunacy" will become common sense and the only defensible position. One hopes sufficient Australians will reach that conclusion before climate change reaches the point of no return.

Most Australians never see the coal trains headed to our ports. For many of those who do, the economic benefits are the first thing to come to mind – certainly that's how I felt not so long ago. With each passing year, more of us connect the coal-laden trains with environmental harm. We understand that for every large bite taken from the earth to fill those trains, a very harmful belch occurs somewhere else in the world. But the connection is vague, we don't really grasp the extent of the damage, and we feel powerless to do much about it. It is only a matter of time before this changes. Today, when I see coal carriages and their cargo pass by, I think of the factories that they fuel; and then of the results of their ever-increasing emissions: the hundreds of species that each large coal-fired power station threatens with extinction, the millions of people at risk from flood, hurricane, fire and drought, and a natural world which seems itself under threat. I am left wondering how many of us will remain unwitting conscripts in the coal war, and how many will become conscientious objectors.

In his speech on election night in 2007, Kevin Rudd acknowledged the courageous work of Bernie Banton, who campaigned so long and hard against the procrastination of those responsible for the trade in asbestos. Like coal, asbestos still causes great harm today. Like coal, asbestos is still being mined in large quantities, increasingly so in China. Yet no one talks about "asbestos leakage" to China. To Australia's credit it did not wait for the rest of the world to get out of asbestos. The question today is whether Australia will keep turning a blind eye to an even bigger threat from its precious quarry, one that will kill far more people and do vastly more harm. The decades between exposure to asbestos and the onset of deadly mesothelioma provided industry and government with an ideal excuse for delay; with coal, the distance and decades between cause and effect provide even more excuses. It will be to Australia's everlasting shame if we treat this threat differently merely because we cannot yet invoke the names of its victims, because the vested interests in this case are more powerful, or because the representatives of those interests have promised us a benign future.

SOURCES

1 "a national prime-time audience": "Nelson maintains stance on emissions trading scheme," *7.30 Report*, ABC TV, 30 July 2008. In 2006–07, the resources sector accounted for 12.9 per cent of GDP, including mining (coal, oil and gas, metal-ore mining, other mining, services to mining) and related manufacturing (processing of metals, non-metal mineral products, petroleum, coal and chemicals). Source: Department of Industry, Science and Research.

2 "be burnt cleanly": see, for example, "Beattie calls for support on clean coal," *AAP*, 4 February 2007; "Anna Bligh pushes coal as Kevin Rudd faces the consequences," *The Australian*, 16 July 2008.

4 Geoffrey Blainey, *The Rush That Never Ended*, Melbourne University Press, Melbourne, Fifth Edition, 2003.

4 G. Blainey, *The Rush*, op. cit., p. 1.

5 "settled in Melbourne." Russel Ward, *The Australian Legend*, Oxford University Press, Melbourne, Illustrated Edition, 1978, p. 138.

5 "a risky proposition for foreign investors": G. Blainey, *The Rush*, op. cit., p. 100.

5 "William Spence rose to become": "William Guthrie Spence: Nation Builder, Labour Leader, Worker," tribute by Bill Shorten at the opening of the new AWU Melbourne headquarters (named after Spence), 1 October 2003.

6 "the nerve-centre of Australian mining": on Collins House, see "Can Australia Survive the Twenty-First Century?" Wilfred Brookes Memorial Lecture, speech by Hugh Morgan at Deakin University, 25 April 2006.

6 "The chairman of BHP, Harold Darling": Robert Macklin and Peter Thompson, *The Big Fella: The Rise and Rise of BHP Billiton*, Random House, (forthcoming 2009).

6 "and nation-builders": Sir Herbert Gepp was another larger-than-life Collins House figure appointed to a senior public role by a conservative leader (Stanley Bruce). For more on the influence of Essington Lewis, see G. Blainey, *The Steelmaster: A Life of Essington Lewis*, Melbourne University Press, Melbourne, 1996.

6 "conflicts of interest": for more on the Mungana Affair and discussion of the Comalco Affair, see K.H. Kennedy, *The Mungana Affair: State Mining and Political Corruption in the 1920s*, University of Queensland Press, St Lucia, 1978, pp. 119–20.

7 "Gordon Nuttall accepted from coal magnate Ken Talbot": "Talbot, Nuttall defend payment," *The Courier Mail*, 8 February 2008; see also Clive Hamilton, *Running from the Storm*, UNSW Press, Sydney, 2001, p. 74.

7 "Russel Ward argues in": R. Ward, *Legend*, op. cit., pp. 169–70.

7 "Vance Palmer drew heavily on": Guy D. Pearse, *An elucidation of the sources of Vance*

Palmer's mining and political novel, Golconda, *particularly in relation to its central character, Macy Donovan*, BA Honours Thesis, James Cook University, 1989.

8 "led to rioting and violence": R. Ward, *Legend*, op. cit., pp. 161–3.

8 "racism may well be the *only* major addition to the Australian character": R. Ward, *Legend*, op. cit., pp. 157, 176.

8 "found themselves speared": G. Blainey, *The Rush*, op. cit., p. 96.

8 "Aboriginal Australians were summarily executed": G. Blainey, *The Rush*, op. cit., p. 96.

9 "even the church turned a blind eye to mining-share speculation": G. Blainey, *The Rush*, op. cit., p. 99.

9 "the Lucky Country": Donald Horne, *The Lucky Country*, Penguin, Melbourne, Fifth Edition, 2005.

10 "undoing some of the old controls and supports": D. Horne, *Lucky Country*, op. cit., from Introduction to the Fifth Edition.

10 "breaking out in different ways in different places": from Correspondence by George Seddon on "Beautiful Lies" by Tim Flannery, in *Quarterly Essay 10*, Black Inc., Melbourne, 2003, p. 104.

11 "around 11 per cent": *Key Facts Australian Industry* 2006–07, Department of Innovation, Industry, Science and Research. Available at <www.innovation.gov.au/Section/Industry/Documents/KeyfactsAustralianIndustry200607.pdf>.

11 "Quiet Boom: How the long economic upswing is changing Australia and its place in the world," John Edwards, Lowy Institute Paper 14, Sydney, 2006, p. 93.

11 "doubling mining's share of GDP in 2007–08 to around 8 per cent": ABS 5204.0, Industry Share of GVA, *Australian System of National Accounts*, 31 October 2008.

11 "car manufacturing was to the United States in its 1950s heyday": Tom McCarthy, *Auto Mania: Cars, Consumers and the Environment*, Yale University Press, New Haven, 2007, p. 101.

11 "more than 75 per cent of GDP": *Key Facts Australian Industry* 2006–07, Department of Innovation, Industry, Science and Research. For more on the contributions of services to GDP, see ABS 5204.0, op. cit., 2008; see also "An Introduction to the Australian Economy," paper by Saul Eslake, Chief Economist, ANZ Bank, Fourth Edition, January 2007.

11–12 "around 10–20 per cent of our export income": in 2007, coal exports were worth $21 billion out of exports valued at $217 billion. In 2007–08, the value of coal exports more than doubled with a spike in coal prices. With the recent fall in coal prices, the value of coal exports as a share of total exports is set to fall in 2009–10.

12 “around 3–4 per cent”: for more on Australia’s relatively low dependence on exports, see Craig James, “China’s massive economic stimulus, the fading glory of the US economy,” in *Economic Insights*, CommSec, 10 November 2008, p. 3.

12 “more people work in cafes and restaurants than in mines”: the mining industry in 2007–08 employed 135,000 people; cafes and restaurants employed 145,000. See *Cafes, Restaurants and Catering Services* 2006–07, ABS 8655.0, 28 April 2008, available at <www.abs.gov.au/AUSSTATS/abs@.nsf/mf/8655.0>; and *Key Facts Australian Industry* 2006–07, op. cit.

12 The aluminium industry employs close to 18,000 including contractors; Bunnings employs 26,000 people. See Australian Aluminium Council and Bunnings websites.

13 “Foreign companies are even more dominant in LNG”: the North West Shelf and the Bayu Undan project in the Timor Sea are almost entirely owned by foreign multinationals.

13 “foreign firms are lining up”: Alex Mitchell, “Iemma’s power privatisation to favour Chinese state giant,” *Crikey.com*, 28 February 2008.

13 “Australian-sounding businesses are anything but”: “Fond farewells for Richard Cottee at Queensland Gas,” *The Australian*, 28 November 2008; and “BG announces friendly takeover bid for Queensland Gas,” *Herald Tribune*, 29 October 2008.

13 “the largest ‘producer’ of coal rather than the largest ‘exporter’”: Dennis Shanahan and Patrick Walters, for example, said in early 2008: “Australia is the world’s largest coal producer and China the world’s largest coal consumer” in “Rudd’s Climate Warning,” *The Australian*, 12 April 2008.

13 “enthusiastically embraced by the Rudd government”: “Minister outlines Australia’s energy future,” media release by Martin Ferguson MP, 5 June 2008.

14 “net energy importer by value”: *The Heat Is On: the Future of Energy in Australia*, report by the Energy Futures Forum, CSIRO, Canberra, 2006, pp. 16, 30; see also *Total Primary Energy Production – All Countries*, US Energy Information Administration, available at <www.eia.doe.gov/pub/international/iealf/tablef1.xls>.

14 “only 3.1 per cent of revenue”: “Mini-Budget 2008–09 New South Wales Government,” 11 November 2008, pp. 4-2 and 4-5.

14 “royalties are based on iron ore and LNG, not coal”: “2008–09 Budget: Economic and Fiscal Outlook,” Government of Western Australia, Budget Paper No. 3, 8 May 2008.

14 “$18 billion in state and federal taxes”: *The Australian Minerals Industry and the Australian Economy*, Fact sheet by the Minerals Council of Australia, November 2008.

15 “taxpayer provides some $9 billion annually”: Chris Riedy, “Energy and Transport Subsidies in Australia,” report by the Institute for Sustainable Futures, University of Technology, Sydney, 2007, p. iv. Available at <ww.cana.net.au/documents/ISFsubsidiesreport2007.pdf>.

15 “a new brown-coal-fired power station”: source “Q&A: The proposed HRL coal-fired power station: Why are the Victorian and Federal Governments bankrolling a polluting coal-fired power station?” Fact sheet from Environment Victoria, 2008. Available at <www.envict.org.au>.

15 “Coal Transport Infrastructure Investment Program”: sources “State Budget 2008–09, Budget Highlights,” Queensland Government, June 2008; “Budget Strategy and Outlook 2008–09,” Queensland Government, June 2008, p. 99. Available at <www.budget.qld.gov.au>.

15 “22 per cent of the ASX total”: sources “Resources sector – One third of ASX trading,” *ASX Newsletter*, November 2006; “ASX and Australian Resources” and “Australian Market Overview,” Australian Securities Exchange website, 2008. Available at <www.asx.com.au>.

15 “the value of resources stocks”: resources (mining, metals and energy) rose to 27 per cent of the All Ordinaries Index at the peak of the commodities boom, but by December 2008, it was back at 22.5 per cent. Source: ASX, ibid.

16 “likely duration of boom conditions”: Paul Cleary, “Mining boom could bust us,” *The Age*, 11 November 2007.

16 “once the wells run dry”: Peter Hartcher, “Heed the curse of a lucky country,” *The Sydney Morning Herald*, 9 May 2008.

16 “the windfall was spent on new liabilities.” For a good discussion of how Australia squandered this boom on permanent new financial liabilities, ignoring the likelihood that the boom revenue was temporary, see the comments by Chris Richardson of Access Economics the day after the 2006 Budget: “Economic analyst assesses Budget tax cuts,” *The World Today*, ABC Radio, 10 May 2006.

17 “As John Garnaut recently wrote”: John Garnaut, “Australia’s ride on China’s coat-tails may be over,” *The Age*, 18 October 2008.

17 “stimulate the economy”: in February 2009, the government announced that federal tax revenue would be $115 billion lower than expected between 2008 and 2012 due to the “global recession, dramatic slowing in China and unwinding of the commodity boom.” See “Macroeconomic and Fiscal Outlook,” media release by the prime minister, 3 February 2009.

18 “humans are predominantly responsible for the rise”: see *Climate Change 2007: Synthesis Report, Summary for Policymakers*, United Nations Intergovernmental Panel

on Climate Change, available at www.ipcc.ch.

18 "Northwest and Northeast Passages": Leigh Phillips, "Europe's Arctic adventure – The new cold rush for resources," *EUObserver.com*, 7 November 2008.

18 "return the climate system to safe territory": see James Hansen, "The Perfect Storm," presentation to the Royal College of Physicians, London, January 2008. Available at <www.columbia.edu/~jeh1/2008/RoyalCollPhyscns_20080129.pdf>.

19 "catastrophic tipping points": for more on why 350ppm and how it might be reached, see James Hansen et al., "Target Atmospheric CO_2: Where should humanity aim?" 2008. Available at <www.columbia.edu/~jeh1/2008/TargetCO2_20080407.pdf>; see also D. Spratt and P. Sutton, *Climate Code Red*, Scribe Publications, Melbourne, 2008; Andrew Glikson, "21st Century Climate Tipping Points," *Opednews.com*, 21 November 2008; Durwood Zaelke et al., "Abrupt Climate Changes Approaching Faster than Previously Predicted: Fast-Track Climate Mitigation Strategies Needed," *MEA Bulletin*, Institute for Governance and Sustainable Development, November 2008.

19 "four metres a century": Ross Garnaut, *Garnaut Climate Change Review*, Cambridge University Press, Melbourne, p. 94.

19 "According to Sir Nicholas Stern's 2006 assessment": Nicholas Stern, *Stern Review: The Economics of Climate Change, Summary of Conclusions*, 2006, p. 1, available at <www.hm-treasury.gov.uk/d/CLOSED_SHORT_executive_summary.pdf>.

19 "Australia has a big stake": see Department of Climate Change; Garnaut *Climate Change Review*; The Climate Institute; and CSIRO.

20 "national leaders come together": see, for example, "G8 plans 50 per cent reduction in greenhouse gases," *The Independent*, 8 July 2008; "G8 backs climate change science, sets no hard goals," *CNN Europe*, 8 June 2007.

20 "environment minister Robert Hill once noted": Ian Lowe, *Living in the Hothouse: How Global Warming Affects Australia*, Scribe Publications, Melbourne, 2005, p. 193.

21 "production has already peaked in more than fifty countries": Kjell Aleklett, "The oil supply tsunami alert," *Energy Bulletin*, April 2005.

21 "60 per cent self-sufficient": *Yearbook Australia 2007*, Australian Bureau of Statistics, Canberra, p. 480; see also *Australian Production of Oil and Gas, Key Statistics*, APPEA, 2007.

22 "petrol prices of up to $8 a litre": *Fuel for Thought: The Future of Transport Fuels – Challenges and Opportunities*, CSIRO, Canberra, June 2008, p. 10.

22 "our overall economy is not especially energy-intensive." As Ross Garnaut noted, "Australia's economy is the eighth most energy intensive among OECD countries. It is about 5 per cent less energy intensive than the world average

and about 8 per cent more energy intensive than the OECD average." *Garnaut Climate Change Review*, op. cit., p. 158.

22 "15 per cent of exports": source *Carbon Pollution Reduction Scheme: Australia's Low Pollution Future*, Green Paper, Commonwealth Government, July 2008, p. 313.

23 "deep cuts would *delay* the trebling of the economy and doubling of real wages by a few years": this is true of the analysis by, among others, ABARE, Allen Consulting Group and the Federal Treasury. See *Economic Impact of Climate Change Policy: The Role of Technology and Economic Instruments*, ABARE Research Report 06.7, ABARE, July 2006, p. 4; *Deep Cuts in Greenhouse Gas Emissions – Economic, Social and Environmental Impacts for Australia*, Report for The Australian Business Roundtable on Climate Change, Allen Consulting Group, Canberra, March 2006, p. 36; *Australia's Low Pollution Future: The Economics of Climate Change Mitigation*, Commonwealth Treasury, 2008; *An Australian Cost Curve for Greenhouse Gas Reduction*, McKinsey & Company, 2008.

23 "acting sooner generates about a quarter of a million jobs more": on an extra 250,000 jobs, see *Deep Cuts in Greenhouse Gas Emissions*, ibid., 2006, p. 36.

23 "a 10 per cent fall in GDP." For a discussion of how ABARE's projections have been manipulated and details on the sources, see Guy Pearse, *High & Dry: John Howard, Climate Change and the Selling of Australia's Future*, Penguin, Melbourne, 2007, pp. 377, 460.

23 "Great Depression in Australia": Saul Eslake, "What is the difference between a recession and a depression?" *Club Troppo*, 23 November 2008; see also <www.australianhistory.org/great-depression.php>.

23 "Barnaby Joyce slams Garnaut climate recommendations," *Herald Sun*, 22 February 2008.

23 "Malcolm Turnbull surrenders in Coalition row over forest plan," *The Australian*, 4 December 2008.

23 "insiders from both major parties have routinely warned": for a good example, see comments by Grahame Morris (John Howard's former chief of staff) on "Friday Forum, the Week in Politics," *Lateline*, ABC TV, 30 March 2007.

23 "Howard-era claims": sources: letter to Tony Nutt (principal private secretary to Prime Minister Howard) from Harold Clough, personal communication, 7 August 2001; "Submission to the Inquiry into the Kyoto Protocol by the Joint Standing Committee on Treaties," The Lavoisier Group, 28 August 2000.

24 "according to the Department of Climate Change": *Tracking to the Kyoto Target 2007: Australia's Greenhouse Emissions Trends, 1990 to 2008–12 and 2020*, Department of Climate Change, Canberra, 2007, p. 4.

24 "on track to meet its Kyoto target": *Tracking to the Kyoto Target 2007*, ibid., p. 4.

25 "Australia's reliance on coal-fired electricity intensified": to see how the emissions intensity of Australia's primary-energy supply has risen since the early 1970s, while the OECD average has fallen, see Figure 7.7 in *Garnaut Climate Change Review*, op. cit., p. 159.

25 "energy policy blueprint": *Energy 2000: A National Energy Policy Paper*, Department of Primary Industries and Energy, AGPS, Canberra, 1988.

25 "one long-time lobbyist": Guy Pearse, *The Business Response to Climate Change: Case Studies of Australian Interest Groups*, PhD thesis, Australian National University, Canberra, 2005, p. 328.

25 "a senior energy policy adviser": G. Pearse, *Business Response*, ibid., pp. 331–2.

26 "Paul Keating recently confirmed his concern": "Australian Workers Union backs Liberals on emissions cap," *The Australian*, 8 August 2008.

27 "one of the dirtiest electricity systems in the world": for more information on how emission-intensive our electricity system is compared with the OECD and global averages, see Garnaut *Climate Change Review*, op. cit., p. 153–60.

28 "hydro projects in Tibet": "Chinese aluminium giant sets up mining unit in Tibet," *The Economic Times* (India), 5 September 2008; and "China plans string of dams in South Tibet," *China Digital Times*, 14 October 2008.

28 "bio-char is now seen as a potential substitute for coking coal": the CSIRO, for example, is championing the use of charcoal as a substitute for coal in iron- and steel-making. See "Mineral Processing: Green Steel," Julian Cribb, *Solve*, CSIRO, May 2006. See also "Essar to set up 2 million tonne gas-fired steel plant in Bangladesh," *Topnews.org*, 4 October 2008; "Bolivia to decide gas supply to JSPL by December," *Steel Guru*, July 2008; "Chop wood, make steel: Hill Wood Products expands its renewable fuel base to serve US Steel," *Ag Innovation News*, September 2002.

28 "investing billions of dollars elsewhere in renewable energy": Rio Tinto Alcan also uses hydro power at Bell Bay smelter in Tasmania but relies on fossil fuels for the vast majority of its Australian production; "BHP eyes $3.4 billion Congo smelter," *The Australian*, 24 October 2007; and "BHP Billiton eyes Congo's Inga dam complex to power new $3 billion aluminium smelter," Bank Information Center, 14 November 2007.

28 "executive vice-president recently told a European audience": "Alcoa signs agreement to support geothermal power research in Iceland," media release by Alcoa, 11 September 2007.

29 "Barack Obama is a longstanding supporter of the coal industry": Obama told an election-campaign audience in September 2008: "Clean-coal technology is something that can make America energy-independent. This is America – we

figured out how to put a man on the moon in ten years. You can't tell me we can't figure out how to burn coal that we find right here in the United States of America and make it work." The quote has since featured in coal-industry advertising. See "Clean-coal debate pits Al Gore's group against Obama, Peabody," Daniel Whitten, *Bloomberg.com*, 4 February 2009; and <http://greenworldads.blogspot.com/2008/12/barack-obama-clean-coal-commercial-ad.html>; and "A clean coal confrontation," Viveka Novak, *Newsweek*, 23 January 2009. For more on Obama's record, see "Coal and Clear Skies: Obama's Balancing Act – An Investigative Report of Presidential Hopeful Barack Obama's Environmental Philosophy," Ben Whitford, *Plenty Magazine*, 31 October 2008.

32 "bought chairs on an ABARE steering committee": for a full list of the fossil-fuel lobby members that bought spots overseeing ABARE's work, along with the amounts they paid, see Clive Hamilton, *Running from the Storm*, UNSW Press, Sydney, 2001, p. 57; or Australian Senate Hansard, *Questions on Notice*, No. 565, 15 May 1997, p. 3518.

32 "ABARE greenhouse policy work": details on the work commissioned from ABARE by the MCA and ESAA are discussed in G. Pearse, *High & Dry*, op. cit., pp. 221, 437; the original interview material is at G. Pearse, *Business Response*, op. cit., p. 328.

33 "contributed considerable sums towards ABARE": a list of contributors to ABARE's general research budget can be found in the finest of fine print at the very bottom of various ABARE reports. The amounts contributed are not disclosed as far as I can establish. For more detail on the links of some contributors to the carbon lobby, see G. Pearse, *High & Dry*, op. cit., pp. 198, 220–1.

34 "by carbon-lobby cash": for more on ACIL and CRA International, and their government work, see G. Pearse, *High & Dry*, op. cit., pp. 155, 204–8, 255.

34 "deny openly the link between greenhouse-gas emissions and climate change." Some prominent (mostly retired) mining bosses continue to dismiss the science to this day: Hugh Morgan, Arvi Parbo, Harold Clough and John Ralph are perhaps the most obvious examples.

34 "fund front groups to challenge the findings": for more on the carbon lobby's links to the sceptic movement, see G. Pearse, *High & Dry*, op. cit., pp. 199–203, 211–8, 250–4; see also <www.desmogblog.com> and <www.sourcewatch.org>.

34 Most of these claims can be found in *Nine Lies About Global Warming*, Lavoisier Group, 2006. See <www.lavoisier.com.au/articles/greenhouse-science/climate-change/lav2006-forWeb.pdf> and "Remarks at the launch of *Nine Lies*,"

Ray Evans, Parliament House, 11 May 2006, <www.lavoisier.com.au/articles/greenhouse-science/climate-change/ninelieslaunch.pdf>. See also "Submission to the Inquiry into the Kyoto Protocol ..." The Lavoisier Group, op. cit., 2000. The rest can be found in G. Pearse, *High & Dry*, op. cit., pp. 142–75. More information on Ray Evans, *Thank God for Carbon*, Lavoisier Group, 2009, is available at the Lavoisier Group website.

34–35 "funding links between climate sceptics and resources companies": for a good example of the links exposed in the USA, see Ross Gelbspan, *Boiling Point*, Basic Books, New York, 2004.

35 "veterans from previous industry-backed campaigns": for example, see George Monbiot's discussion of the links between climate sceptics Fred Seitz, Fred Singer and Steve Milloy, and the denial campaign run by the tobacco industry in *Heat: How to Stop the Planet Burning*, Allen Lane, London, 2006, pp. 30–36. See also G. Pearse, *High & Dry*, op. cit., pp. 286–7, 442, 448.

35 "funding of Australia's sceptics": see, for example, "Public misperceptions of human-caused climate change: the role of the media," testimony of Dr Robert M. Carter, James Cook University, Townsville, Australia, before the Committee on Environment and Public Works, United States Senate, 6 January 2006; evidence from Dr Ian Castles, Australian National University, Canberra, *The Economics of Climate Change*, House of Lords Select Committee on Economic Affairs, London, 2005, p. 80; to see lots of sceptics travelling at once, see the mother of all sceptic gatherings: the 2008 International Conference on Climate Change (which also featured Bob Carter). For more detail on it and its co-sponsors (which included the IPA), see <www.sourcewatch.org/index.php?title=The_2008_International_Conference_on_Climate_Change#Speakers_at_the_Conference>; the conference program is at <www.heartland.org/NewYork08/PDFs/ConferenceProgram.pdf>.

35 "funds the globe-trotting greenhouse advocacy": see, for example <http://members.iinet.net.au/~glrmc/>.

35 "a sustainability report": *Towards Sustainable Development*, Community Environment Report by Western Mining Corporation, 2000, p. 28; see also "WMC backs new climate sceptics group," *Mining Monitor*, July 2000, p. 4.

35 "funded by emission-intensive companies": the IPA's Sydney-based counterpart, the Centre for Independent Studies, which has also denied the scientific and economic cases for emission cuts but been less active in recent years, was funded into existence by five Australian Industry Greenhouse Network (AIGN) companies, with seed funding organised by the boss of Western Mining Corporation, Hugh Morgan. See G. Pearse, *High & Dry*, op. cit., pp. 199, 243, 440.

35 "they'd stop funding us": G. Pearse, *High & Dry*, op. cit., p. 282; see also *Inquiry into the National Access Regime*, transcript of proceedings, Productivity Commission, Melbourne, 28 May 2001, p. 45. For an example of the lopsided science promoted by the IPA, from which its fossil-energy funders like to keep a comfortable distance, see "Some Australian views on climate change, abstracted by Bob Carter from the published writings of Professor Bob Carter, Mr Ian Castles, Professor Aynsley Kellow, Mr William Kininmonth, Professor Garth Paltridge, Professor Ian Plimer," occasional paper from the Institute of Public Affairs, 28 February 2005, <www.ipa.org.au/library/Carter2004_CLIMATEBROCHURE.pdf>; see also "Science, Statisticians and the Prophets of Doom," Ian Castles, *IPA Review*, December 2001, <www.ipa.org.au/library/Review53-4%20Scientists%20Statisticians%20the%20prophets%20of%20doom.pdf>.

35 "the AIE said": "Is Global Warming Cause for Alarm?" Speech by Bob Carter, Australian Institute of Energy, Melbourne, 27 April 2005 (has been removed from AIE website but author has a copy on file. The invitation can be accessed at <http://ipa.peptolab.com/library/Global%20Warming%20Luncheon%20-%2027%20April%202005.pdf>); see also "Climate Change – Cool Science or Hot Air?" Speech by Bob Carter, AIE in Sydney 2004. For another example of the IPA organising sceptics to speak at public events, see "Doubtful Nats scrutinise climate science," *Stock Journal*, 1 August 2008.

35 "a front group": when the AEF was formed, the office address it registered with ASIC was the same as the IPA, and two of its directors were senior IPA staff. For more on the IPA's role in establishing the AEF, see G. Pearse, *High & Dry*, op. cit., p. 434 and <www.sourcewatch.org/index.php?title=Australian_Environment_Foundation>.

36 "The foundation claims that": "Getaway GetUp on the ETS," media release by the Australian Environment Foundation, 29 December 2008.

36 "the right to veto research findings": regarding CSIRO censorship on behalf of contributors (and the comments about "constant pressure" which come from Ian Smith – former CSIRO assistant chief of the coal and energy technology division, now the energy technology division) see "Old King Coal," Polly Simons, *NewMatilda*, 15 November, 2007, <http://newmatilda.com/2007/11/15/old-king-coal>.

36 "barred from making public comment": see, for example, the comments of Graeme Pearman and Barney Foran in "The Greenhouse Mafia," Janine Cohen, *Four Corners*, ABC TV, 13 February 2006.

37 "spread over more than a decade": the Australian Coal Association announced an extension of the voluntary levy out to 2017 in 2007 and they expect this

to raise a billion dollars. See <www.coal21.com.au/Media/COALFUND11May.pdf>.

37 "permanent alibi": in "What is Rudd's Agenda?", Robert Manne, *The Monthly*, November 2008, p. 32.

37 "renewable technologies 'off the shelf.'" See G. Pearse, *High & Dry*, op. cit., pp. 155–56. See also David Montgomery and Anne Smith, "Workshop on the Economics of Climate Change: Understanding Transatlantic Differences," presentation at Resources for the Future Conference, Charles River Associates International (CRA), Washington DC, 2 March 2006.

38 "an alliance representing almost all of Australia's biggest fossil-energy producers and consumers": for a full list of AIGN members, go to <www.aign.net.au/membership>.

38 "greenhouse mafia": for more on the "greenhouse mafia," see G. Pearse, *High & Dry*, op. cit., and "The Greenhouse Mafia," Janine Cohen, *Four Corners*, op. cit.

38 "the Clean Energy Council": for more on how the coal and oil interests dominated the Australian Gas Association, see G. Pearse, *High & Dry*, op. cit., pp. 187–9. In relation to the same dynamic at the Clean Energy Council, the owners of three of Victoria's brown-coal-fired power stations are prominent members of the CEC: the chairman Richard McIndoe is drawn from one of these companies. The perceived failure of the CEC to represent the interests of its members is the subject of legal action. See "Renewable energy body in turmoil," *Herald Sun*, 12 July 2008; "'Dirty' fuel firms split clean energy group," *The Age*, 13 July 2008; and "Clean-up row heads for court," *Herald Sun*, 12 July 2008.

39 "donation was again the largest": this excludes the donations received from John Curtin House Ltd and Labor Holdings Pty Ltd, which are holding companies owned by the Labor Party. The think-tanks affiliated with the Liberal and Labor parties have also received generous contributions from the likes of BHP Billiton and Woodside, and various associated consultants. For more detail on contributions from the carbon lobby, see G. Pearse, *High & Dry*, op. cit., pp. 195–99; and the database available for public search at <http://fadar.aec.gov.au/>; for the CFMEU's self-description, see <www.cfmeu.com.au>.

40 "invited former ALP environment minister Ros Kelly onto its board": Leighton Holdings is also a significant and regular corporate contributor to the ALP.

40 "its chief executive": Ross Rolfe is also a former CEO of the coal-fired generator Stanwell Corporation.

40 "subtly tightens its grip on the country": for a good summary of the connections between the carbon lobby and the Queensland Labor Party, see "Connected," Darryl Passmore, *Sunday Mail*, 17 August 2008.

40 "opposite front benches": Andrew Robb was hired by a consortium including Rio Tinto and BHP Billiton to oppose Kyoto and emissions trading just prior to his election to parliament in 2004. For more on Robb's sceptic views and work for the carbon lobby, see "The Player," Angus Grigg, *Australian Financial Review Magazine*, 29 September 2006.

41 "a game of musical chairs": for more on the extent of musical chairs, see G. Pearse, *High & Dry*, op. cit., pp. 228–38.

41 "dominated by mining company executives": G. Pearse, *High & Dry*, op. cit., pp. 197-8, 240. Greg Gailey, current BCA president, is a long-time oil industry executive and former chairman of the Minerals Council of Australia. For a full list of BCA presidents, see <ww.bca.com.au/Content/100828.aspx>.

41 "dominated almost every greenhouse-related consultative committee": for more on the various government processes dominated by the AIGN (ABARE; CSIRO; the Low Emissions Technology Advisory Group; the Greenhouse Challenge; the Asia Pacific Partnership on Clean Development and Climate; and many more, see G. Pearse, *High & Dry*, op. cit., pp. 218–26, 261–4.

41 "Its leaders dismiss the image": "Carbon Price Signals and Emissions Trading," speech by John Daley, chief executive of the AIGN, Business Council for Sustainable Energy Conference, Brisbane, May 2006.

42 "the official would soon be the lobbyist": for example, Australia's former ambassador for the environment, Ralph Hillman, is now chief executive of the Australian Coal Association; his predecessor as ambassador for the environment, Meg McDonald, became a lobbyist for Alcoa. For more on the many other former government members of Australia's official delegation that the AIGN has recruited, see G. Pearse, *High & Dry*, op. cit. pp. 227–38.

42 "Sometimes the carbon lobby looks lucky": for more on John Daley, who was chief executive of AIGN when he was seconded into PM&C to help the Howard government's Task Group on Emissions Trading, see "Politics, not policy, behind PM's switch," Richard Baker, *The Age*, 19 February 2007; and G. Pearse, *High & Dry*, op. cit. pp. 275–6. For more on the LETAG, see the leaked notes of Rio Tinto's Sam Walsh at <www.tai.org.au/documents/downloads/WP56.pdf>; for more on Howard's consultation with polluters before rolling Cabinet, see G. Pearse, ibid, pp. 86, 422; for more on the sceptic conversion of CRC's association boss (and former Liberal Party president) Tony Staley, see G. Pearse, ibid., pp. 147, 282–4; John Ralph was a member of the panel overseeing the Low Emissions Technology Demonstration Fund (grants from which overwhelmingly favoured the fossil-energy sector). While serving in this position, he said: "I don't doubt the climate is changing, but I don't know

human activity is the primary cause of it." For more, see "Boss queries climate action," *The Australian*, 26 October 2006 and G. Pearse, ibid., p. 145; for more on the Rio Tinto executive that served as Australia's chief scientist, see: G. Pearse, ibid., p 174.

42 "Sometimes lightning even strikes twice": for more on the CEC's new chief executive (who has previously called Al Gore's *An Inconvenient Truth* "a horror film with a contrived plot," see G. Pearse, *High & Dry*, op. cit., pp. 160, 211, 247; see also Clive Hamilton, *Scorcher: The Dirty Politics of Climate Change*, Black Inc., Melbourne, 2007, pp. 189, 198, 200, 233; and <www.sourcewatch.org/index.php?title=Matthew_Warren> and "Announcement of CEO Appointment," media release by the Clean Energy Council, 1 October 2008. In another stroke of apparent luck for the carbon lobby, the Chair of the Clean Energy Council has publicly argued for coal-fired power stations to be compensated for the impact of emissions trading. See "'Dirty' fuel firms split Clean Energy Group," *The Age*, 13 July 2008.

43 "they dismiss climate change as": "Greenhouse sceptics to congregate," *The Age*, 28 February 2007; and "Remarks at the launch of *Nine Lies*," Ray Evans, op. cit.

44 "leader who said Australia should never even have got involved": G. Pearse, *High & Dry*, op. cit., p. 139; see also "The Hot Debate," Liz Jackson, *Four Corners*, ABC TV, 18 August 1997.

46 "Sydney Declaration at APEC in 2007": "APEC – Muriel's Wedding all over again," Guy Pearse, *The Manic Times*, 15 September 2007.

46 "the leaked Turnbull plea": address by Dr John Hewson to the Institution of Engineers Australia, Southern Highlands branch, Mittagong RSL, 27 November 2008; see also "Turnbull and PM at loggerheads on Kyoto," *The Sydney Morning Herald*, 28 October 2008.

48 "they could instead say": comments from Greg Evans, Policy and Economics Director, Australian Chamber of Commerce and Industry, PM, ABC Radio, transcript of interview, 30 September 2008.

48 "'iron triangle' of polluter, political and bureaucratic elites": for more on the concept of "iron triangles" and interest-group theory, see F.R. Baumgartner and B.L. Leech, *Basic Interests: The Importance of Groups in Politics and in Political Science*, Princeton University Press, Princeton, 1998.

49 "a polluter-friendly emissions trading plan": for more on the emissions trading scheme design proposed by Labor's National Emissions Trading Taskforce, see <www.emissionstrading.net.au/>.

49 "Labor might eventually come around to nuclear power": "Labor faces inside push on nuclear energy," *The Australian*, 27 June 2008.

49 "Sceptics could expect a sympathetic ear": Peter Walsh acknowledged Martin Ferguson's participation in the Parliament House launch of the Lavoisier Group's *Nine Facts about Climate Change*. See *President's Report – 2007*, The Lavoisier Group, Melbourne, 2007, <www.lavoisier.com.au/articles/climate-policy/science-and-policy/walsh2007-30.php>.

49 "AWU patriarch Bill Ludwig": Ludwig would later publicly dismiss Ross Garnaut as a "wacko," claim that the climate was not warming, and suggest plugging volcanoes would be more effective than emissions trading in combating climate change. "Climate change adviser branded a wacko by AWU President, Bill Ludwig," *The Australian*, 4 February 2009.

50 "Ferguson reportedly told": this is attributed to Ferguson by Louis Hissink (a scientist who attended the meeting) in his comments on 3 March 2007 on the blog run by the IPA's Jennifer Marohasy. See <www.jennifermarohasy.com/blog/archives/001922.html>.

50 "as Rudd put it": "Rudd in the hot seat": *7.30 Report*, ABC TV, 11 December 2008.

50 "Nelson's opposition would be as dominated by greenhouse denial": Nelson appointed two veterans of the delay campaign, including Peter Hendy from the Australian Chamber of Commerce and Industry, and Tom Switzer from *The Australian*. Since Nelson's demise, Tom Switzer has gone to work for the IPA. For more on his sceptic-friendly work at *The Australian*, see C. Hamilton, *Scorcher*, op. cit., pp. 199–200. For more on Hendy, see G. Pearse, *High & Dry*, op. cit.

50 For more on the Howard-era sceptics, many of whom remain in powerful positions in the post-Howard Opposition, see G. Pearse, *High & Dry*, op. cit. pp. 142–7.

50 "Nelson's successor would later acknowledge this": in January 2009, Turnbull said: "The Coalition's position on this issue [climate change and emissions trading] is very well known – it's the same that we had in government." See "No split on emissions scheme, Turnbull," *ABC News Online*, 14 January 2009.

51 "since the collapse of communism": "The Player," Angus Grigg, *Australian Financial Review Magazine*, 29 September 2006.

51 "sceptic as chief of staff": a renowned sceptic, Chris Kenny, was appointed as Turnbull's chief of staff in late 2008. See "Turnbull's chief of staff Chris Kenny hit by taser for cameras," *The Australian*, 8 January 2009; "Malcolm's strange new spindoctor," Greg Barns, *Crikey.com*, 9 January 2009. Another adviser hired by Turnbull was Alex Robson who, prior to his appointment, publicly denied carbon dioxide was a pollutant, called scientific consensus an "irrelevant concept"

and dismissed emissions trading as fashionable because it allowed socialists to hijack the tools of the free market. See "The biggest hoax ever perpetrated on the Australian public," Alex Robson, *The Australian*, 20 November 2006; and "Leftist critics often get it wrong," Alex Robson, *The Australian*, 12 December 2006.

51 "how to spend emissions trading revenue": see "Henry Ergas' review of the Commonwealth, State, Territory and Local tax system," media release by Malcolm Turnbull, at <www.ergasreview.com>. On letting the planet warm, see "Climate cure more costly than disease," Henry Ergas, *The Australian*, 14 July 2008. His associate at Concept Economics, Brian Fisher, had previously said it might be more efficient to let Pacific Islands go under than save them by cutting emissions: C. Hamilton, *Running from the Storm*, op. cit., p. 79.

51 "National Party, meanwhile, elected Barnaby Joyce": for Barnaby Joyce's suggestion that pressure from "environmental goose-steppers" (i.e. Nazis) is becoming scary, see "Nationals climate mutiny looms for Turnbull," *The Australian*, 14 January 2009; for more on his views on mining Antarctica before others get to it, see "Antarctica ripe for mining," says Joyce, *ABC News Online*, 1 May 2006.

51 "Turnbull finally announced his climate-change policy": "A Green Carbon Initiative – Investing in our Land, Energy and Food Security and Jobs," speech by Malcolm Turnbull at the National Convention of the Young Liberal Movement, 24 January 2009.

52 "a new climate change religion." See "Beware deep green luddites on climate," editorial, *The Australian*, 8 June 2007; and "Let the great debate on climate change continue," editorial, *The Australian*, 14 February 2007.

53 "AIGN's ranks now included": BHP Billiton is no longer listed by the AIGN as an individual company member, but it is still represented through the Australian Coal Association and the Minerals Council of Australia.

53 "emissions trading could plunge Australia into darkness": "Compo urged for power stations," *The Australian*, 23 February 2008; "Big Australian power generators won't give up," Olga Galacho, *Herald Sun*, 12 July 2008; "Garnaut has faith in market forces," *The Australian*, 27 March 2008.

54 "price of electricity would rise": "Power plants in danger from emissions tradings (sic) scheme," *The Australian*, 25 July 2008; "States' power play pulls plug on ETS," *The Australian*, 2 July 2008; "Power producers warn on emission targets," *The Australian*, 24 May 2008.

54 "gas prices might double": "Gas price under pressure," *The Australian*, 1 July 2008.

54 "the same consultants": CRA International was commissioned by the Australian Petroleum Production and Exploration Association. See *Implications of a 20 per cent renewable energy target for electricity generation*, report for APPEA by CRA International (Authors: Anna Matysek and Brian Fisher), Canberra, November 2007, p. 4.

54 "similar proportion of their income on energy": this was confirmed to me in an email from Brian Fisher (then executive director of ABARE, later consultant for CRA International). He acknowledged that real wages were projected to grow by 81 per cent even if Australia pursued deep-cut scenarios. Personal communication, 24 July 2006. See also G. Pearse, *High & Dry*, op. cit., p. 378.

54 "Back in Kyoto in 1997": for more on Australia's negotiating position in Kyoto, see G. Pearse, *High & Dry*, op. cit., pp. 14, 73–4.

54 "could not have been reasonably anticipated": these comments came from Brad Page, CEO of the Energy Supply Association of Australia. See "No mercy for dirty power says Garnaut's climate report," *The Australian*, 4 July 2008.

54 "Business Council of Australia claimed": "New research shows a way forward on emissions trading," media release by the Business Council of Australia, 21 August 2008; "Carbon plan a company killer," *The Australian*, 22 August 2008.

55 "said the Australian Industry Group": "Fears climate policy to cost a million jobs," Andrew Fraser, *The Canberra Times*, 7 October 2008.

55 "the industry body recommended a 'dry run'": "Better cut carbon dioxide later," Heather Ridout, *The Australian*, 11 December 2008.

55 "according to the AWU's Paul Howes": "Australian Workers Union backs Liberals on emissions cap," Lenore Taylor, *The Australian*, 8 August 2008.

55 "would leak offshore, said Howes": "Indulgent Greens miss point of White Paper," Paul Howes, *ABC News Online*, 19 December 2008.

55 "entire industries would shrink, with commensurate job losses": "Treasury releases modelling on emissions trading scheme," *7.30 Report*, ABC TV, 30 October 2008. For a good example of ABARE's pioneering work in this kind of misrepresentation, see "The Economic Impacts of Uniform Emission Abatement," Brian Fisher and Stephen Brown, presentation to the Countdown to Kyoto Conference, Canberra, 19–21 August 1997, p. 11.

55 "What they concealed": the AIGN suggested the following percentage declines in output by 2050 relative to an undisclosed business-as-usual scenario: a 75 per cent fall for non-ferrous metals like aluminium, agriculture 45 per cent less, steel 53 per cent smaller, coal mining falls by 33 per cent, and electricity and other energy by 25 and 60 per cent respectively. The impression created could hardly be in greater contrast to Treasury's modelling, which found that

compared with today the following changes in output would occur by 2050: iron ore (+234 per cent); other mining (+120 per cent); cement (+106 per cent); non-ferrous mining (+93 per cent); alumina (+73 per cent); metal products (+54 per cent); even coal mining (+66 per cent). Sources: "Energy State of the Nation," presentation by Mike Hitchens, chief executive of the Australian Industry Greenhouse Network, 18 March 2008, p. 6; *Australia's Low Pollution Future: The Economics of Climate Change Mitigation*, op. cit., p. 164.

56 "What if Woodside followed through on its threat": "Woodside floats plans for pirate plant," *The Sydney Morning Herald*, 14 November 2008.

56 "actually meant what they said?": "DJ Nyrstar: Australia carbon trading scheme could shut smelters," *Trading Markets.com*, 12 November 2008; "Labor divided over emissions trading," *The Sydney Morning Herald*, 13 November 2008; "Sun Metals threatens offshore move if ETS goes ahead," *The World Today*, ABC Radio, 13 November 2008.

57 "around 94 per cent of GDP would survive": Figure 9.2 in *Carbon Pollution Reduction Scheme*, Green Paper, op. cit., p. 313.

57 "As for LNG, industry analysts believe": "Woodside doth protest too much," Giles Parkinson, *Business Spectator*, 26 September 2008.

57 "ABARE found in 2007": *Report of the Task Group on Emissions Trading*, Australian Government – Prime Ministerial Task Group on Emissions Trading, PM&C, 2007, p. 172.

57 "Treasury's 2008 modelling": *Australia's Low Pollution Future: the Economics of Climate Change Mitigation*, op. cit., pp. 3, 170, 195.

58 "In his final report, Garnaut said": *Garnaut Climate Change Review*, op. cit., p. 45.

59 *Garnaut Climate Change Review*, op. cit., p. xxx; see also "Garnaut doubtful of global action," *Lateline*, ABC TV, 5 September 2008.

60 "The biggest polluters should have almost all their emission costs paid": for more detail of the compensation proposed by Garnaut for emission-intensive trade-exposed industries, see pp. xxxii, 341–50 of the final Garnaut report, and pp. 38–40 of the *Garnaut Emissions Trading Scheme Discussion Paper*, March 2008.

60 "It's also a cheap option": in many developing countries, the cost at which it makes more financial sense to protect rather than clear the forest equates to less than US$1 per tonne of CO_2. One Indonesian study in 2008 estimated that most of the carbon dioxide lost through forest and peatland conversion generated less than US$5 a tonne; 40 per cent generated less than US$1 a tonne. See "Less than $1.00 per ton of CO_2 – Research suggests Indonesia can reduce emissions with sustainable benefits," Centre for International Forestry Research, 30 July 2008; see also Garnaut report, op. cit., pp. xxv, 235, 238.

60 "Companies like Rio are eagerly eyeing similar deals": "Mining company signs carbon trading deal with graziers," media release, Carbonconservation.com, 17 November 2006.

61 "access to low-cost abatement options": *Garnaut Climate Change Review*, op. cit., p. 238.

61 "Matt Warren at the *Australian* summed up the strategy": "Proposal blind to political realities," Matt Warren, *The Australian*, 22 February 2008.

62 "close to fifteen dollars more being spent subsidising fossil-energy use": source: Treasury and C. Riedy, "Energy and Transport ..." op. cit., 2007.

62 "the government dug deeper": "Minister outlines Australia's energy future," media release by Martin Ferguson MP, 5 June 2008; and "Turn blowtorch on APEC: Rudd," *Brisbane Times*, 8 June 2008.

62 "he told audiences in both countries": speaking in Jakarta, Rudd said, "The carbon credits from avoided deforestation – in other words the preservation of rainforests – could be an alternative source of revenue for Indonesia." In PNG, he said much the same: "Our work on forest carbon has the potential to generate substantial investment flows into PNG." See "Australia and Indonesia – inseparable partners working together and working together in the world," speech by Kevin Rudd, Jakarta, 13 June 2008; and "Address to the PNG business and alumni breakfast," speech by Kevin Rudd, Port Moresby, 7 March 2008.

63 "compensated for the decline in the value of their businesses": *Carbon Pollution Reduction Scheme*, Green Paper, op. cit., pp. 340–90.

63 "lethargic sectors in the greenhouse debate": Department of Climate Change; ABS; G. Pearse, *High & Dry*, op. cit., p. 330.

63 "the dirtier the industry, the bigger the compo": the Green Paper proposed that those sectors generating 1500–2000 tonnes of CO_2-e per million dollars in revenue would receive 60 per cent of their permits free; those generating even higher levels would get 90 per cent.

63 "paying for one in every five tonnes of the carbon dioxide": under the criteria set out by the Rudd government's Green Paper, industries responsible for 24.8 per cent of emissions qualify for 90 per cent of permits free (excluding electricity generation); and industries accounting for 12.5 per cent of emissions would qualify to receive 60 per cent of permits free. Across the two categories, the industries might pay for an average of 20 per cent of emissions. The agricultural sectors that meet the threshold would pay for none of their emissions until agriculture is included in the scheme. See *Carbon Pollution Reduction Scheme*, Green Paper, op. cit., pp. 313, 321, 338 and 498.

63 "Rudd's energy minister said": see the Rudd government's enthusiasm for coal to liquids, for example, in "Australia's energy security and the clean energy challenge," speech by Martin Ferguson, 5 June 2008.

63 "former coal-industry lobbyist was tasked with": "Low Emission Coal Initiatives Announced," media release by Martin Ferguson, Minister for Resources and Energy, 28 July 2008.

64 "a new Global Carbon Capture and Storage Institute": see "Global Carbon Capture and Storage Initiative," joint media release by Kevin Rudd and Martin Ferguson, 19 September 2008.

64 "the 'non-assisted' might pay three times as much": ABARE's very relevant finding was buried towards the back of the report eventually released. See *Report of the Task Group on Emissions Trading*, op. cit., pp. 172, 176.

64 "prices up and destroying communities": "ETS would raise food prices, lower exports," *The Australian*, 10 January 2009; "Senators unite over carbon sink forests tax," *The World Today*, ABC Radio, 26 June 2008; "Nationals support food on kitchen table over carbon sinks," media release by Senator Barnaby Joyce, 1 December 2008.

65 "the government believes": the Treasury has projected that if Australia adopts a 5 per cent emission-reduction target for 2020 (which it has), this might be associated with additional carbon-sink forest plantations of around 5 million hectares. The industry projections I have seen suggest a much more modest expansion. Between 2005 and 2050, Treasury's analysis suggests that another 5 million hectares of carbon sinks might generate emission savings of about 29 million tonnes a year (including around 6.7 million tonnes already anticipated annually). The "40 million hectare" figure used recently alongside the claim that land in south-east Queensland is under threat relates to Treasury modelling of Garnaut's 25 per cent cut by 2020 scenario, which is not being pursued by the government. Source: *Australia's Low Pollution Future: the Economics of Climate Change Mitigation*, op. cit., pp. 185–6.

65 "make the credits fully fungible well before 2020": *Garnaut Climate Change Review*, op. cit., pp. 337–41; *Carbon Pollution Reduction Scheme*, Green Paper, op. cit., pp. 240–1.

65 "accounting for some 1750 million tonnes of CO_2 per year": the estimate for the CO_2 from deforestation in Indonesia and PNG comes from Ross Garnaut, *Garnaut Climate Change Review – Draft Report*, Commonwealth of Australia, Canberra, June 2008, p. 329.

67 "take on reductions comparable to Australia": from *Carbon Pollution Reduction Scheme*, White Paper, op. cit., vol 1, p. iv.

67 “with a straight face Kevin Rudd said”: “Australia’s Low Pollution Future: Launch of Australian Government’s White Paper on the Carbon Pollution Reduction Scheme,” speech by the prime minister, National Press Club, Canberra, 15 December 2008.

68 “according to Treasury’s projections”: Treasury projected 585 million tonnes for 2020, but this includes land-use change and forestry emissions, which the Department of Climate Change estimates at 24 million tonnes in 2020. See *Tracking the Kyoto Target 2007*, Department of Climate Change, Canberra, pp. 13–14.

68 “on a per capita basis”: under the proposed 5 per cent target, emissions per capita (excluding land clearing and forestry) improve by around 9.1 per cent, not the 34 per cent claimed by Rudd; under the 15 per cent target, there is an 18.2 per cent per capita improvement, not the 41 per cent claimed by Rudd. Sources: Department of Climate Change; Commonwealth Treasury.

68 “twice as large as the EU’s by 2020”: “End the economic mismanagement of the national interest,” media release by the Climate Institute, 16 December 2008.

69 “‘Nobody gets a free ride,’ said Penny Wong”: “Protest increases over polluters’ discount,” *The Sydney Morning Herald*, 17 December 2008.

69 “A glossy pamphlet”: *The Carbon Pollution Reduction Scheme and You: Creating a Low Pollution Economy for the Future*, Commonwealth Treasury, Canberra, 2008, p. 2.

69 “polluters would pay for only a tiny fraction”: the reduced threshold for companies receiving 60 per cent of their permits free means more compensated companies are paying for 40 per cent of their emissions. On average, however, those who meet the 60 and 90 per cent compensation thresholds would pay for around 22.5 per cent of their emissions.

69 “so a $750 million abatement program”: “Warm up act in climate war,” *The Australian*, 20 December 2008.

70 “All households would also ‘benefit’”: *Scheme Impact on the Cost of Living*, Fact Sheet, Australian Government, December, 2008. available at <www.climatechange.gov.au/whitepaper/factsheets/pubs/026-scheme-impact-on-the-cost%20of-living.pdf>.

70 “Deciphering what cheques were coming”: the re-jigged payments to householders include, among others, Family Tax Benefit (Parts A and B), the Newstart Allowance, the Low Income Tax Offset, the Dependent Spouse Tax Offset, Parenting Payment, Age Pension, Seniors Concession Allowance, Youth Allowance, Widow Allowance, Sickness Allowance, Mature Age Allowance, Partner Allowance, Special Benefit, Crisis Payment, Exceptional Circumstances Relief

Payment, Austudy and ABSTUDY, Disability Support Pension, Carer Payment, Parenting Payment (single and partnered), Bereavement Allowance, Widow Class B Pension, Wife Pension, and the Veterans' Affairs Service Pension and Disability Pension.

71 "no mention of the one-off $93 cheque being written": "Warm-up act in the climate war," *The Australian*, 20 December 2008.

71 "on average $500 per household": "Householders to foot the big polluters' carbon bill," media release by the Australian Conservation Foundation, December 2008.

71 "Rio Tinto would get over half a billion dollars annually": for more on which companies are likely to receive cheques of free permits worth hundreds of millions of dollars (in a few cases, it runs into the billions), see *Research Note: The impact of industry assistance measures under the Carbon Pollution Reduction Scheme*, Innovest: Stategic Value Advisers, 18 December 2008.

72 "They made their obligatory protests": "Greenpeace, WWF damn paper on climate change," *The Sydney Morning Herald*, 16 December 2008; interview with Ralph Hillman on the *Morning Show with Ross Solly*, ABC Radio 666, Canberra, 16 December 2008; and also "More money needed for ETS: energy industry," *The Sydney Morning Herald*, 2 February 2009.

72 "Ross Garnaut attempted to distance himself": "Carbon plan fuels meltdown," *The Sydney Morning Herald*, 20 December 2008.

73 "Paul Kelly's take on Rudd's policy": "Diabolically clever politics," Paul Kelly, *The Australian*, 17 December 2008.

73 "As Janet Albrechtsen wrote": "Blessed change in the climate," Janet Albrechtsen, *The Australian*, 17 December 2008.

74 "Labor seems intent on accommodating them all": "Scramble for carbon trading exemptions," *The Australian*, 3 May 2008.

74 "The man who said so often in Opposition": "Federal government's export strategy not working: Rudd," *AAP General News*, 3 March 2006; "Rudd unveils new education policy," *7.30 Report*, ABC TV, 23 January 2007.

74 "'huge fiscal churn'": "Diabolically clever politics," Paul Kelly, *The Australian*, 17 December 2008.

75 "endless war against nature": from "Silent Country: travels through a recovering landscape," Tim Winton, *The Monthly*, October 2008.

75 "Extended producer responsibility": EPR is a concept gaining increasing acceptance across a wide range of environment policy issues, particularly packaging and waste management more generally. It recognises that industry responsibility does not end at the factory door or with the sale of the product.

Instead, industry bears a responsibility for environmental damage done through the use of its products and the environmental costs should be incorporated into the cost of the product.

75 "a major source of domestic emissions": coal burning for electricity generation results in around 200 million tonnes of CO_2 emissions annually in Australia. In addition to CO_2 emissions, coal mining generates significant quantities of methane, a potent greenhouse gas.

77 "switching to green power": for more information on the availability of green power, see <www.greenpower.gov.au>.

77 "exports to 2030 are projected to be": Australian LNG exports are projected to rise to between 67 and 76 million tonnes a year by 2030. Emissions associated with its use are approximately 187–212 million tonnes. This is between 15.8–17.9 per cent of the 1182 million tonnes of CO_2 likely to be associated with the 438 million tonnes of annual coal exports projected for 2030. Sources: ABARE; US Energy Information Administration; "LNG Survey: LNG's hot future," *The Diplomat*, 22 August 2007; *Australian Energy – National and State Projections*, ABARE Research Report 07-24, Australian Bureau of Agricultural and Resource Economics, Canberra, 2007, p. 5; conversion factors courtesy of Energy Strategies Ltd.

77 "The best summary of the situation, by James Hansen": for more on the importance of coal in the emissions equation, see J. Hansen et al., *Target Atmospheric CO_2 – Where Should Humanity Aim?* op. cit.; and A. Glickson, "Twenty-first century climate tipping points," op. cit.

78 "In a recent letter to Barack Obama": "Open Letter to Barack and Michelle Obama from James and Anniek Hansen," 29 December 2008, available at <www.columbia.edu/~jeh1/mailings/20081229_DearMichelleAndBarack.pdf>.

78 "annual emissions equivalent of 25,000 cars": on 8 October 2008, two vessels leaving Dalrymple Bay Coal Terminal loaded 247,792 tonnes of coal – a new daily record. Assuming an average annual emission of 5 tonnes per car, the coal loaded onto each of these vessels equates to the annual contribution of around 25,000 cars. Source: Babcock and Brown Infrastructure.

79 "our coal exports are projected to double": sources: *US Energy Information Administration – International Energy Outlook* 2008, at <www.eia.doe.gov/oiaf/ieo/pdf/table8_mst.pdf>; and *Australian Energy – National and State Projections* 2029–30, ABARE, at <www.abareconomics.com/interactive/energy_dec06/htm/chptr_4.htm>.

79 "as the prime minister says": "Rudd's $100m climate institute," *The Sydney Morning Herald*, 19 September 2008.

79 “100 tonnes of CO_2 actually captured and stored”: this assumes 100,000 tonnes of CO_2 captured and/or stored annually, which is roughly what CCS demonstration projects in Australia amount to in the next few years, if all goes well. Source: CO_2 CRC.

79 “blocks of coal with electric cords”: to see the electricity cord plugged straight into a chunk of coal, see <www.stopgreenwash.org/coal>.

80 “the slogan ‘I believe’ seventeen times”: to view the “I Believe” ad on You Tube, see <http://au.youtube.com/watch?v=X_5OrJVR_Vc>.

80 “‘Flick the switch’ and ‘Turn it down’”: these slogans were used in ads placed online in the climate section of *The Australian* in December 2008 – same pie-in-the-sky blue, naturally. The author holds copies on file.

80 “the Kevin07 campaign”: according to media reports, the NewGenCoal campaign for the ACA was based on research and advice from ad-man Neil Lawrence and Labor pollster John Utting. Sources: “New generation coal technologies for the next generation,” media release by the NSW Minerals Council, 12 November 2008; “Energy revolution vital,” Marian Wilkinson, *The Sydney Morning Herald*, 13 November 2008.

80 “the coal industry’s NewGenCoal website”: to view the upbeat video messages from CSIRO employees on the coal industry’s NewGenCoal PR website, see <http://newgencoal.com.au/randdtv.aspx>; the government’s view that there is “no alternative” to more coal-fired power stations comes from Martin Ferguson: “Ferguson backs coal-fired power,” *The Australian*, 3 April 2008. The name change – dropping “clean coal” from the government’s own CCS flagship program – was mentioned in parliament by Lindsay Tanner in December 2008: *House of Representatives Hansard*, Second Reading, Appropriations Bill (No.3), 4 December 2008.

80–81 “was abandoned”: see “Chimneys sweep BP clean coal plan away,” Nigel Wilson, *The Australian*, 10 May 2008.

81 “also shelved recently”: “Shell, Anglo rethink local coal to liquids project,” *Reuters*, 2 December 2008.

81 “might save 100,000 tonnes of CO_2 annually”: this excludes projects that have not publicly estimated their tonnes per annum sequestration plans and larger projects like Santos’ “carbon storage hub” at Moomba and Woodside’s Gorgon project, which are primarily about enhanced oil extraction and/or have nothing much to do with coal-fired power. In the case of Moomba, the proponents have expressed doubts about whether it would be viable to take CO_2 from third parties (like coal-fired power stations) for less than $50–70 a tonne. Source: Cooperative Research Centre for Greenhouse Gas Technologies, <www.co2crc.

com.au/demo/>; "Carbon capture serves the national interest: Ferguson," *The Australian*, 21 November 2008.

81 "one carbon-lobby-funded research centre": CRC for Greenhouse Gas Technologies, ibid.

81 "another Chinese CCS project": "Carbon capture milestone for CSIRO in China," CSIRO media release, 31 July 2008 at <www.csiro.au/news/CarbonCaptureMilestone.html>; and "PM visits pilot clean-coal energy plant in China," *The World Today*, ABC Radio, 11 April 2008.

81 "International estimates suggest": presentation at the Aker Clean Carbon CCS Seminar and Opening Ceremony in Norway by Rachel Crisp, deputy director, Cleaner Fossil Fuels Unit, UK Department of Energy and Climate Change, 16 October 2008.

81 "carbon price of A$100 a tonne": "Coal industry reaches out for love," Tim Colebatch, *The Age*, 13 November 2008.

81 "the IPCC doesn't expect": *IPCC Special Report on Carbon Dioxide Capture and Storage – Summary for Policymakers, A Special Report of Working Group III of the Intergovernmental Panel on Climate Change*, 2005, p. 13.

81 "Modelling by the Australian Treasury suggests": *Australia's Low Pollution Future: The Economics of Climate Change Mitigation*, op. cit., p. 179.

82 "John Boshier, told the ABC's *7.30 Report*": "Going underground: carbon dioxide storage experiment," *7.30 Report*, ABC TV, 7 April 2008.

82 "finally built with CCS": *The True Cost of Coal*, Greenpeace International, November 2008.

82 "a power-company executive from Stanwell Corporation": testimony of Mr Howard Morrison, manager, Emerging Technologies, Stanwell Corporation Ltd, Committee Hansard, Inquiry into Geosequestration Technology by the House of Representatives Committee on Science and Innovation, 11 September 2006, p. S&I 17.

83 "the 'Rolls-Royce' of carbon capture and storage": Tim Flannery in discussion with Robert Manne, *Slow TV*, November 2008, available at <www.themonthly.com.au/tm/node/1310>.

83 "senior officials at the International Energy Agency say": "Coal's future is safe, but what about the climate?" *ABC News Online*, 5 August 2008.

84 "The prime minister murmurs": "Rudd Defends Emissions Trading Scheme," *7.30 Report*, ABC TV, 30 October 2008.

84 "emissions are to peak globally by 2015": on the need for this, see "Open Letter to the Prime Minister of Australia from Australia's climate-science community," 26 September 2008, available at <www.science.unsw.edu.au/ruddletter/>.

84 "skin in the game": "Australia funds carbon capture institute," *The Sydney Morning Herald*, 19 September 2008; "fair and reasonable" comes from Ralph Hillman, CEO of the Australian Coal Association – see "Doubts raised over Ross Garnaut's emission targets," *The Australian*, 2 October 2008.

84 "is expected to fetch in 2008–09": this is based on US$300 a tonne for hard coking coal and US$100 a tonne for thermal coal in the year 2008–09, the current split between thermal and coking coal, and an exchange rate of $A=US$0.62. Sources for the price estimates come from the ABS, Queensland Treasury and ABARE.

85 "expanding railway infrastructure to double coal export capacity": the Rudd government's National Low Emissions Coal Initiative is worth $500 million. On 12 December 2008, the prime minister announced that "$580 million of today's investment will be used to expand capacity along the rail corridors connecting Hunter Valley coal mines to the port of Newcastle. This $1 billion project will more than double the amount of coal being transported to export markets from 97 to 200 million tonnes a year." See "$4.7 billion nation building package," media statement by the prime minister, 12 December 2008. Other sources: Budget 2008–09, Budget Paper Number 2, Part 2: Expense Measures (Climate Change), Table 3 Tackling Climate Change, available at <www.budget.gov.au/2008-09/content/bp2/html/expense-06.htm>.

85 "as much greenhouse pollution annually as Australia's national total": around 200 million tonnes of additional coal will be exported from the Hunter and Glandstone regions as a result of the new infrastructure funded or approved by the Rudd government. This is roughly equivalent to adding 540 million tonnes of CO_2. Sources: "Gladstone port expansion passes critical environmental milestone," joint statement by Anna Bligh and Paul Lucas, 21 April 2008; "Australia – world's climate vandal," *Green Left Weekly*, 18 January 2009.

85 "just one large clean-coal power station installed by 2020": the Australian Coal Association estimated that its 10,000 GWh target was equivalent to 1500 MW (or one large coal-fired power station). See "Historic alliance calls for a national task force on carbon capture and storage," joint media release from the Australian Coal Association, the Climate Institute, WWF and the CFMEU (Mining & Energy), 16 April 2008.

86 "Terry McCrann says": "Held to ransom by cut-price exports," Terry McCrann, *The Australian*, 6 December 2008.

87 "China has also subsidised petroleum": for a useful discussion of energy subsidies in China, see Usha C.V. Haley, *Shedding light on energy subsidies in China: An analysis of China's steel industry from 2000–2007*, Alliance for American Manufactur-

ing, Washington DC, January 2008; See also "Fuel subsidies for some make oil more expensive for all," *International Herald Tribune*, 28 July 2008.

87 "an extra 400 million people": "Has China's one child policy worked?" *BBC News*, 20 September 2007.

87 "by exporting around 8 billion tonnes of coal": this assumes average coal exports of 350 million tonnes per year between now and 2030 at an average coal price of A$250 per tonne. In 2008–09, exports are projected to earn $43 billion, which is roughly A$165 per tonne. So this is likely to be a very conservative estimate.

88 "twice as many people work in McDonald's": McDonald's claims to employ 75,000 Australians; coal mining employs approximately 30,000 people. See <www.mcdonalds.com.au/HTML/inside/company.asp> and <www.mcdonalds.com.au/PDFs/CSR_Report_2007.pdf>.

88 "one to two years to recover every cent of coal-export revenue": this is based on estimated coal-export income of $43 billion annually and a $1.1 trillion economy that generates around $21 billion a week. Sources: ABARE, Treasury.

88 "Employment would recover, too": Ian Lowe, *Quarterly Essay 27: Reaction Time*, Black Inc., Melbourne, 2007, p. 10.

88 "with Australia accounting for more than half of that growth": "International Energy Outlook 2008," Table 8, US Energy Information Administration, at <www.eia.doe.gov/oiaf/ieo/pdf/table8_mst.pdf>.

88 Saudi net oil exports are currently around 8.7 million barrels a day, which equates to 3.175 billion barrels a year. This equates to around 1.36 billion tonnes of CO_2 annually. Australia's coal exports are projected to reach 438 million tonnes a year by 2029–30. This equates to 1.182 billion tonnes of CO_2 annually – roughly 87 per cent of the current carbon footprint of Saudi net oil exports. (Conversion factor: 430 kilograms of CO_2 per barrel of crude oil. Source: USEPA.)

89 "Resorting to the dope-dealer's defence": for more on the use of the "drug-dealer's defence" by mining companies in relation to uranium, see Ian Lowe's Correspondence to "Reaction Time" in *Quarterly Essay* 28, Black Inc., Melbourne, 2007, p. 123.

89 "Mick Keelty, recently said": Inaugural Ray Whitrod Oration, by Commissioner Mick Keelty, Adelaide Convention Centre, 24 September 2007.

91 "polluted a host of rivers already": for more information on how coal mining is destroying waterways in eastern Australia, see <www.riverssos.com/news.html>. See also the leaked report of May 2008 from the NSW Department of Environment and Climate Change, <www.riverssos.com/pdf_files/

deccreport.pdf>, pp. 25–27.

92 "produced by diversified multinationals": BHP Billiton, Rio Tinto, Anglo Coal and Xstrata reportedly account for 75 per cent of black-coal production. Among the remaining 25 per cent, many companies are also diversified. Sources: CFMEU and Rocklands Richfields Ltd.

92–93 "350 million Chinese still likely to move": "Australia's ride on China's coattails may be over," John Garnaut, *The Age*, 18 October 2008.

93 "IEA have underestimated the potential of renewable alternatives": "Wind power in context – a clean energy revolution in the energy sector," Energy Watch Group, Basel, December 2008; and "International Energy Agency 'blocking global switch to renewables,'" *The Guardian*, 9 January 2009.

94 "Tim Flannery stated in late 2008": Tim Flannery in discussion with Robert Manne, *Slow TV*, op. cit.

95 "threatens with extinction": climate change already kills some 150,000 people annually according to estimates from the World Health Organization. See "Climate shift tied to 150,000 fatalities," *The Washington Post*, 17 November 2005; and "Climate and Health – Fact Sheet," World Health Organization, 2005. For more on the number of species threatened with extinction per coal-fired power station, see "Open Letter to Barack and Michelle Obama from James and Anniek Hansen," 29 December 2008, op. cit.

AMERICAN REVOLUTION | *Correspondence*

Gavin Kitching

If there is a good place to be in a global financial meltdown, New York is clearly it. You get a sense of the crisis as a human product: if you actually know or live next door to some of the people involved, it is easier to see through the insistent media attempt to present it as some kind of uncontrollable natural disaster – a "tsunami" or a "meltdown." Even the terms "crash" and "disaster" have this effect. It becomes altogether easier to see it as the unintended product of the actions of fallible human beings if you are walking your dog alongside them.

The journal form, and the way Kate Jennings' excitement and trepidation over the election clearly mirrored that of hundreds of people around her and online, made our own obsessions seem okay. There was an almost hysterical feeling that Obama was somehow America's last chance, perhaps even the world's last chance, and if the great US public somehow managed to screw it up again, not only would the world go to hell in a handcart, but the devil would chortle all the way. It may not seem like this at all in three or four years' time. (After all, I remember a similar kind of euphoria when Tony Blair was elected in Britain in '96, after seventeen years of Tory rule.)

Of course Obama will disappoint in certain respects: I expect that, whatever he does, the Arab–Israeli conflict will still be going on; some initiatives won't work or will misfire; there'll be a corruption incident or three. And then we'll all be wise and judicious and balanced, and start giving six out of ten. But still, he's *got* to be better than Dubya! The troops will surely leave Iraq. His administration will start to treat the rest of the world with some degree of respect. He will have to do something about US infrastructure and, in doing so, create work for people. And he won't be afraid to tighten the regulatory grip on Wall Street with an enforcement agency that is both properly staffed and equipped.

Jennings' essay has finally helped me grasp the problem – I mean the essential problem – behind the financial crisis, and it isn't what I expected. When it began,

I found myself struggling to understand what was happening. I thought that there were people who understood the problem and that if I read enough financial columns and articles I'd grasp it. And of course that's not entirely wrong. Nouriel Roubini had some insight ahead of time (unlike Felix Rohatyn, for example, who just did the usual reluctant after-the-fact stuff). But this approach obscured the heart of the problem: the disconnection of the hundreds of thousands (even millions) of people just doing their thing in front of a computer screen from the combined effect and outcome of their actions. What *American Revolution* reveals, better than anything else I've read on the topic, is that most bankers and their trading minions haven't got the faintest idea "what's gone wrong" or why; that they were dealing in mechanisms they didn't understand. So when it all unravelled, they were as pathetic and gobsmacked as the rest of us. Hence the generally underwhelming impression they made in front of Senate committees. I note that Jeffrey Sachs was inclined to make Alan Greenspan the evil spider at the centre of the web and one who *did* know what he was doing. Although that's true of his cheap-money policies, even he, on his own admission, did not understand derivatives and assumed that banks had both the information and incentive structure to be rational risk-takers, rather than helpless jockeys on runaway mounts.

Perhaps, in a capitalist system, people do have to be protected from themselves. If the "planned economy" is out because it doesn't work, then paternalism (and maternalism) is in. People have to be told that they *can't* do certain things in the marketplace, because while it might be good for them in the short run, it isn't in the general interest in the long run. And those "certain things" can be everything from zero-deposit mortgages, to "leverage levels" of fifty and above, to the clever-clever "securitisation" of risk which, when aggregated, produces massive, insecure, unpriceable risk.

But I think that's going to be a hard row to hoe. There's a heady "freedom" ethic in capitalism, and especially in American capitalism, which makes that kind of paternalism (especially when the state embodies it) a total ideological anathema. And once the pain of this recession is gone, you can bet your bottom dollar that the rhetoric of "freedom" and "choice" and "individual liberty" versus "Big Brotherdom" will be wheeled out once again, either to try and roll back re-regulation or to make it ineffectual in practice. Keynes is capitalism for grown-ups, but god knows whether there are, or will ever be, enough grown-ups out there to make *social* self-interest real and keep what the old man himself called "animal spirits" in check.

Gavin Kitching

Christina Thompson

Kate Jennings captures perfectly the intensity of these past months, the terrible anxiety we felt, the almost pathological conviction that the Republicans would do anything, say anything, pull out all the stops, and that the Democrats would just stand there like numbskulls while the election was stolen from them once again. It was absolutely excruciating, despite the obvious steely competence of the Obama campaign, and the steady march toward victory that, in hindsight, this clearly was. No matter how fervently we were exhorted to hope, what many of us were experiencing was actually fear.

America has been living in a state of fear – pumped up by the Bush administration whenever it showed signs of flagging – for nearly eight years. Fear of enemies, fear of weapons, fear of difference, fear of science, fear of tolerance, fear of truth. How we can have been so oppressed by our own freely elected government is one for the historians (though Jennings' observations about Cheney's Cheney might be a good place to start). But, in addition to the recent warping of our national mindset, there was our unspeakable history and the particular way it intersected with Obama's campaign. Sometime in the summer my friend Hazel Rowley, who also lives in New York, wrote to me: "We are all thinking about it, aren't we? There's something so saint-like about him and his message of hope. Every time that's bobbed up in America, it's been gunned down." This, of course, was on everybody's mind and the flare-ups of ugliness at some of Sarah Palin's rallies did nothing to reassure us.

So, there were some obvious reasons to be afraid. But there were also some that were not so obvious. The enormous outpouring of joy and relief on the eve of the election – absolutely palpable all over the country (well, maybe not in the south) – suggests that what we were really afraid of was something about ourselves. The election of Barack Obama showed us, as much as it proved to the rest of the world, that we were not as racist a country as we were afraid we might actually be. Our fear of our own dark side – the idea that our worst history might be inescapable – was far more scary and debilitating than increased taxes or the idea that jihadists were coming to murder us in our beds. We were afraid that, when put to the test, America would reveal itself to be, first and foremost, a country of bigots. And the relief that this was not true swept over us like a January thaw.

In the aftermath of the election, many people have been at pains to remind us that America is still a country in which the playing field is not level. All the social indicators – life expectancy, infant mortality, income, educational attainment, incarceration rates – are worse for African-Americans, and none of that changed overnight on 4 November. But the symbolic effect of having a not-simply-white person in the White House is also inescapable. As Frank Rich put it in the *New York Times*, America is not yet post-racial. But, in some sense, Barack Obama is.

An inauguration-day story in the *Times* by Jodi Kantor laid this out in some detail in a description of Obama's extended family. There was the step-grandmother from Kenya, the Indonesian-American half-sister, her Chinese-Canadian husband, and the descendents of both white Americans and African slaves who speak, among them, at least nine languages, including Cantonese, Hebrew and Swahili. This complexity, which must at first have looked like a liability, was in fact a huge advantage for Barack Obama because his opponents were never able to resolve it into a single, clearly alien and scary thing, as they might have been able to do if he were simply African-American.

The fact that while Obama is both African and American he is not actually African-American (that is, not descended, unless on his mother's side, from African slaves brought to North America) also seems to have relaxed the relationship between him and the white people he needed to vote for him if he was going to be elected president of the United States. This is not *why* they voted for him in the end. They did that because, ironically, there were suddenly some new and pressing reasons for terror, namely global economic meltdown, and they wanted someone smart, calm and competent at the helm. But the ambiguities of his identity seem to have made it easier for them to put aside the racial issue, to forget about it for a while and get beyond the diabolical two-step in which white and black people in America have been trapped for hundreds of years. Americans, both white and black, voted for Barack Obama because he is reassuring and encouraging, bold and yet flexible, insightful but firm. He is obviously born to lead and people of all walks and stations of life recognised that. But the way he stayed above the racial fray, accepting it as one more facet of our history to be faced squarely but without histrionics, was hugely liberating for us all. Suddenly we could see a world in which race was important but no longer defining. Suddenly, there were other things to think about – and we did.

Jennings writes that with a week to go before the election a famous UK journalist predicted there would be dancing in the streets. "But I dismiss the optimism as coming from a Brit who doesn't understand race in this country," she writes. "Because I come from an Australian farming background, I'm more pragmatist than pessimist, although if asked whether a glass is half full or half empty, I will answer, 'What glass?'" I had to laugh when I read this. You can live a long time in another country but this is one of the things that will never change. It's one of the things I liked most about this essay: the dry Australian point of view, infused, but only sporadically, with American zeal. Americans are notoriously un-nuanced – earnest, eager, unselfconscious, even slightly silly when viewed from bleaker, wittier corners of the world. But this was another thing about the election that struck me. Our national tendency toward innocence was for a moment elevated from the ridiculous to the sublime.

At the time, I took my own anxiety about the election to be a form of realism – there were lots of legitimate reasons to be worried and afraid – but looking back on it now I wonder if it wasn't rather an absence of hope. "Hope" was, of course, the watchword of the Obama campaign, and I have to confess that I found it slightly irritating. It seemed – how else can I put it? – sort of corny. "Hope" did not strike me as a political message; it was not like "Change" or "Action" or "Restoration of the Public Trust." "Hope" seemed naive, almost simplistic, and too close by half to "Faith." If there is one thing we do not need any more of in America, it's religion in the public sphere. But it seemed clear from the outset that no matter how religious Obama might be in private, he was never going to impose his religious views on the rest of us. This was not someone who believed, as both George W. Bush and Sarah Palin seemed to, that God Himself had chosen him for a mission. Which made Obama's unwavering conviction even more impressive; there was something, okay, *audacious* about his hope.

We progressives in America are like wounded lovers: we've been let down so many times that it's almost impossible for us to believe our time will come. But in the aftermath of an election that I wanted so desperately but could never actually believe would happen, an election that President Obama himself seems never to have doubted, I'm beginning to wonder if there isn't something to that cheerful, can-do attitude for which Americans are so often mocked. The essence of this moment seems to be that after nearly a decade of being ground down, something like optimism really did rise up in the American people. And what's really weird is that it reached its peak just as everything else around us was going straight to hell. In the October that Jennings chronicles, there had never been more reason to be scared and worried, but the push, the shove really, toward pessimism was counterbalanced by an even more powerful force. So nobody did jump. The storyline was too compelling. I wonder if Jennings could be convinced to keep typing? We're all still waiting to find out how it ends.

Christina Thompson

Gavin Kitching is professor of politics at the University of New South Wales and the author of, most recently, *The Trouble with Theory.*

Guy Pearse is a former member of the Liberal Party and was a speechwriter for former environment minister Robert Hill. He has also been an industry lobbyist, consultant and spin doctor. In 2007 he exposed the politics behind Australia's response to climate change on *Four Corners* and in his book *High & Dry.*

Christina Thompson is the editor of *Harvard Review* and the author of *Come on Shore and We Will Kill and Eat You All.*

www.ingramcontent.com/pod-product-compliance
Lightning Source LLC
Chambersburg PA
CBHW061224270326
41927CB00025B/3487

* 9 7 8 1 8 6 3 9 5 3 7 5 7 *